CITY UNDERGROUND SPACE
城市地下空间

凤凰空间·华南编辑部 编

江苏凤凰科学技术出版社

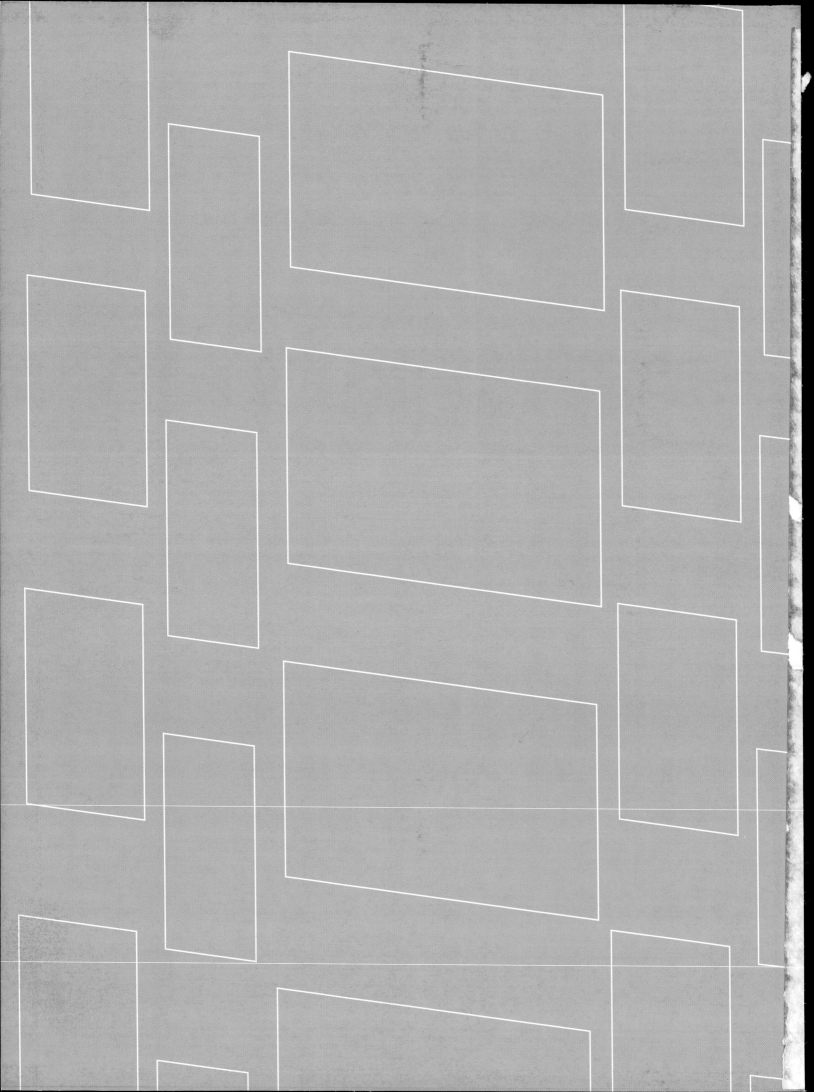

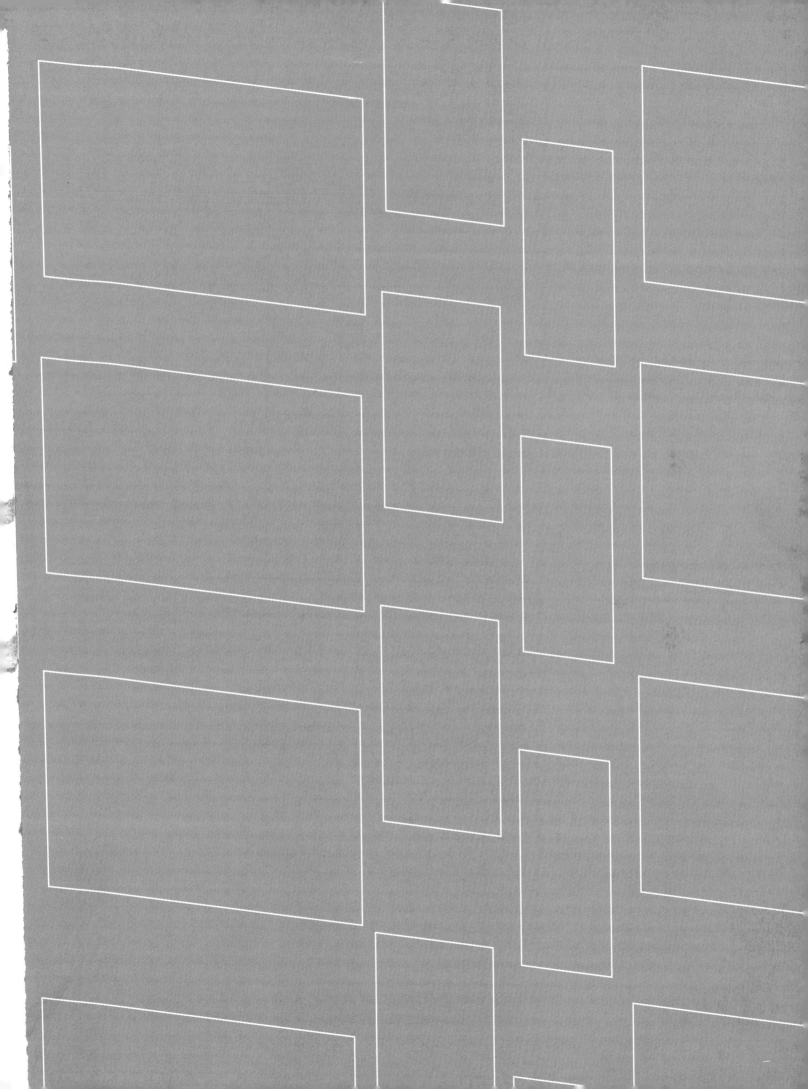

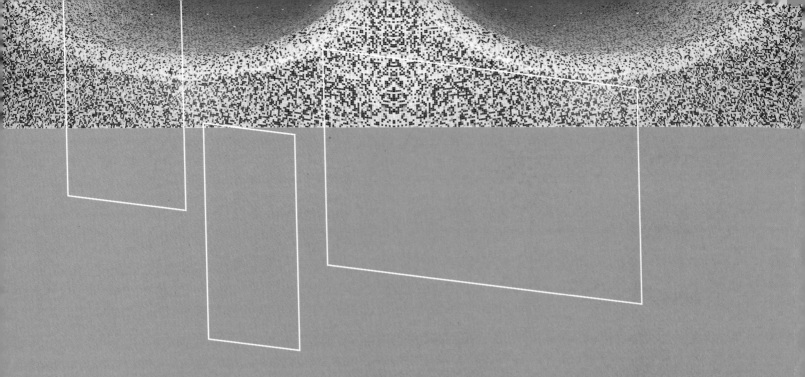

Preface

In the 21th century, people pay more attention to the environmental protection. And the undeveloped above ground spaces are less than before. Therefore, the development of underground space becomes the most important form of the architecture. So far, the development of underground spaces in foreign countries has a long history of 150 years since the London metro was built in 1845. But in China it was started in 1950s, and now it is in the initial stage. The mission of developing underground spaces will be afforded by our generation and those after us.

The book includes more than 28 classic works to show the outstanding underground space planning and design. With innovative shapes, they are planned with not only commercial function, but also public space. What's more, they meet the demands of the shape and the environmental protection. Some showcase good lighting condition and energy saving system; some reflect the characteristic of culture protection; some show excellent design of people flow; some become the city's unique landscape to give kaleidoscopic visual enjoyment to visitors. The works are all around the world. The design based on the local climate, different topography and culture to make the book become more various and artistic.

Through researching, describing and showcasing, the book collects and analyzes the newest wisdom idea of underground space planning and design, which reflect that the trend of underground space design is suitable for the new life style. With high quality actual photos, the book also features the process of every design with the comprehensive original design materials, including plans, elevations, sections and drawings in order to provide very precious and valuable information.

Last but not the least, Many thanks for the contribution given by the architecture companies and architects to this book!

前言

二十一世纪，人们注重环保，地上空间越来越紧迫，于是，地下空间便成为建筑的重要发展方向。国外自1845年伦敦地铁的开始兴建发展到现在，已有150年之久，但中国城市地下空间的开发利用源于上个世纪50年代，还处于起步阶段，发展地下空间是我们这代人乃至几代人的使命。

本书在全球范围内，选录了28个经典的地下空间规划与设计作品。这些建筑造型新颖，不仅具备了地下空间的商业功能和公共空间功能，还体现了美观、环保的要求。它们或在采光节能方面独具特色，或体现了人文保护的特性，或能最大限度地引导人流，或本身就能成为城市景观，或给予游人丰富多彩的视觉享受。它们遍布世界各地，与气候、地形以及各地的文化差异相适应，展现了设计上的多样性和艺术性。

本书通过对这些作品进行研究、描述和展示，全面分析了地下空间规划与设计的最新理念，反映了适合现代人新生活方式的地下空间设计趋势。除了配以高质量的实景照片以外，还辅以全面的原始设计资料，包括平面图、立面图及剖面图，以及建筑构思过程图，为读者提供了非常宝贵的、具有借鉴价值的信息资料。

在本书的编写过程中，得到了众多建筑设计事务所和建筑设计师的大力支持，在此表示感谢！

CONTENTS 目录

Preface
前言

Chapter One 第 1 章

The Space with Good Lighting Condition and Energy Saving Effect
采光节能的空间

Extension of Staedel Museum 施塔德尔博物馆扩建	012
"Hannah Arendt" Professional High School Underground Extension 汉娜·阿伦特职业学校地下学校扩建	026
M4 Fővám tér - Underground Station, Budapest 布达佩斯 M4 Fővám tér 地铁站	040
La Salut Metro Station La Salut 地铁站	060
Urban Railway Stations S7 S7 线城市地铁站	066

第 2 章　Chapter Two

The Space Based on Culture Protection

人文保护的空间

| The Sammy Ofer Wing, Extension of the National Maritime Museum, London | 078 |
| 伦敦国家海事博物馆扩建——萨米·奥弗翼楼 | |

Gammel Hellerup Gymnasium　096
Gammel Hellerup 体育馆

Transforming King's Cross　112
国王十字火车站改建工程

The Earthscraper　130
摩地大楼

Seminary Reform　144
塔拉戈纳神学院改建

Palau de la Música Reform　150
Palau de la Música 音乐厅改造

Købmagergade　160
Kømagergade 购物步行街

Solar Powered Mosque　166
太阳能清真寺

Te Mirumiru Early Childhood Centre　172
Te Mirumiru 儿童早教中心

Chapter Three 第3章

The Space with the Function of Leading People
引导人流的空间

Local Transport Junction at Europaplatz Graz 格拉茨 Europaplatz 广场交通枢纽	184
Danish National Maritime Museum 丹麦国家海事博物馆	208
Vendsyssel Museum of Art 海宁 Vendsyssel 艺术博物馆	230
ON-A's New Office ON-A 建筑事务所新办公室	242
Station 20 Station 20 地铁站	248

Chapter Four 第4章

The Space with Natural Landscape and Man-made Marvels
自然造景的空间

Europa City Europa City 活动中心	256
Ingolftorg, Reykjavik 雷克雅未克 Ingolftorg 广场	268
Masdar Plaza Centre 马斯达生态广场	276
The Cave, New Underground Auditorium of the Tate Modern 泰特现代美术博物馆的新地下礼堂 The Cave	284
Cottages at Fallingwater 流水别墅上的屋子	292

第 5 章　Chapter Five

The Space with Visual Extension Effect
延伸视觉的空间

The Blue Planet "蓝色星球"水族馆	302
San Andreu Metro Station Reform San Andreu 地铁站改造	312
Drassanes Metro Station Reform Drassanes 地铁站改造	320
Toledo Metro Station Toledo 地铁站	332
Index 索引	

Chapter One
第 1 章

Natural light is an effective way to provide lighting for the space below ground and save energy. It also meets the psychological requirements of perceiving nature sunlight, sense of spatial direction, the alternation of day and night, vagaries, seasons, climate, etc. Generally Speaking, Underground space collects nature light by two methods. One is passive lighting, the other is active lighting. Passive lighting gets sunlight from different kinds of windows. And active lighting transmits nature light to very position by using collecting, transmitting, scattering devices with control systems. These two methods can be found in the cases of this chapter. Through artful designs, architects showcase fantastic interior of underground space.

The Space with Good Lighting Condition and Energy Saving Effect

采光节能的空间

对于地下建筑来说，实现自然采光不仅仅是为了满足照度和节约采光能耗的要求，更在于满足人们感知自然阳光、空间方向感、昼夜交替、阴晴变化、季节气候等自然信息的心理要求。总的来说，地下空间利用自然光的方法主要有被动式采光法和主动式采光法两类：被动式采光法是通过利用不同类型的窗户进行采光的方法；主动式采光法则是利用集光、传光和散光等装置与配套的控制系统将自然光传送到需要照明部位的采光法。在本章的案例中，两种采光法均有涉及。建筑师们通过巧妙的设计，为地下空间营造出良好的室内效果。

- **Architects:** schneider+schumacher
- **Project Architects:** Michael Schumacher, Kai Otto, Till Schneider
- **Developer:** Städelsches Kunstinstitut
- **Photographer:** Norbert Miguletz
- **Gross floor area:** Extension 4,151 m² / Total 24,726 m²
- **Gross volume:** Extension 27,568 m³ / Total 115,535 m³

- 设计公司：schneider+schumacher 建筑设计事务所
- 主创设计师：米歇尔.舒马赫、凯、阿托，堤尔、施耐德
- 投资方：施塔德尔博物馆
- 摄影师：Norbert Miguletz
- 地点：德国法兰克福
- 总建筑面积：扩建：4 151 m² / 整体：24 726 m²
- 总体积：扩建：27 568 m³ / 整体：115 535 m³

施塔德尔博物馆扩建

Extension of Staedel Museum

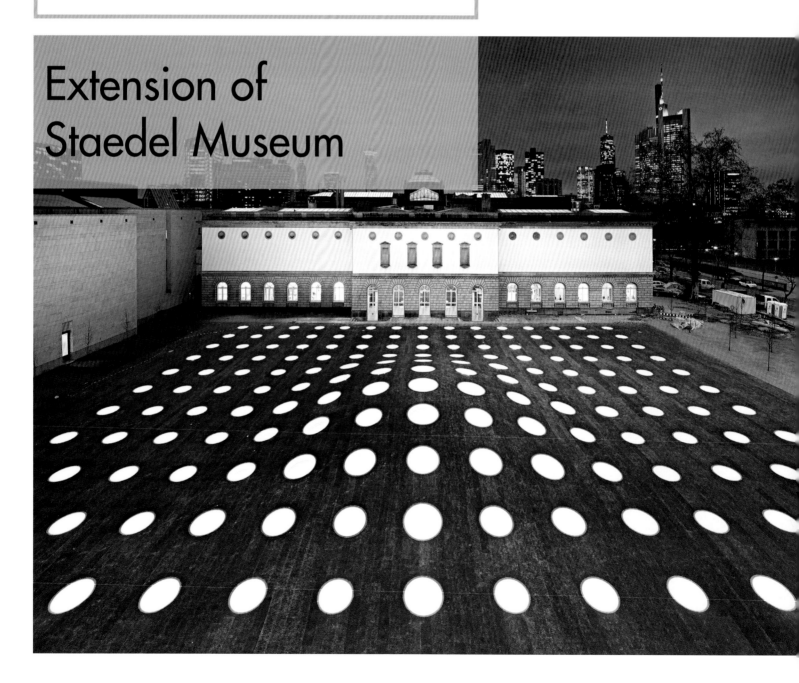

Project Overview

The new building adjoins the garden wing completed at the start of the 20th century and itself the first extension of the original museum building, which was built on Frankfurt's Schaumainkai in 1878.

项目概况

最新加建的施塔德尔博物馆将 20 世纪初完成的花园侧厅与建成于 1878 年的、位于法兰克福美术馆街上的原建筑首次连接起来。

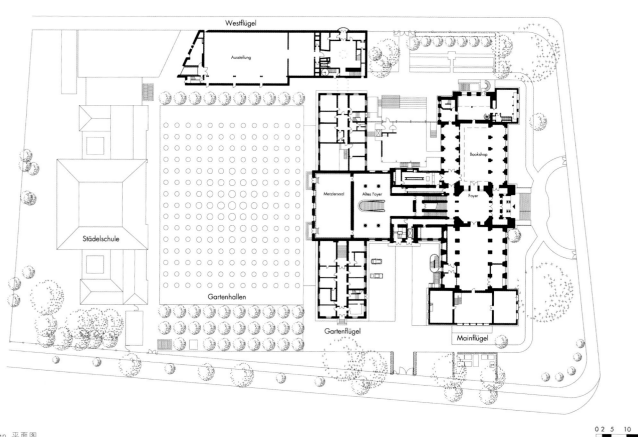

Plan 平面图

In contrast to any of the extension work carried out to date, the new section of the museum will not be above ground; the generous new space planned by schneider+schumacher will be located beneath the Städel garden. The new exhibition space will be accessed via a central axis from the main entrance on the museum's river side. By opening the two tympanums to the right and left of the museum's main entrance foyer, visitors will be able to reach the Metzler Foyer level. A staircase will then lead from this area down into the 3,000 m² museum extension beneath the garden.

与时下的各种加建工程不同，加建的博物馆主体完全坐落于地下，德国 schneider+schumacher 建筑师事务所在原有的施塔德尔花园下规划出了一个十分充裕的建筑空间。新馆的入口处于原展馆的靠近河边的主入口的中心轴线上，从主馆两侧的休息厅进入到达梅茨勒厅，位于达梅茨勒厅后方的楼梯就是新馆的主入口，它将引导你进入地下3 000平方米的展览空间。

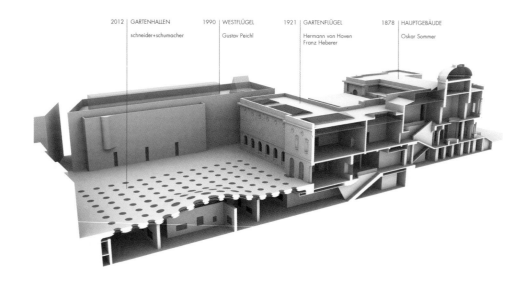

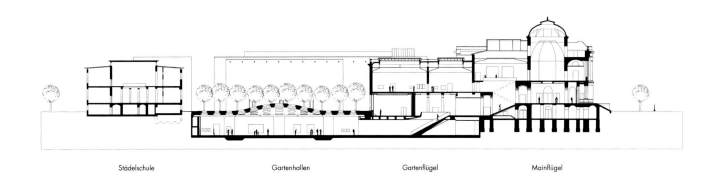

Städelschule   Gartenhallen   Gartenflügel   Mainflügel

0 2 5 10 20

Section 剖面图

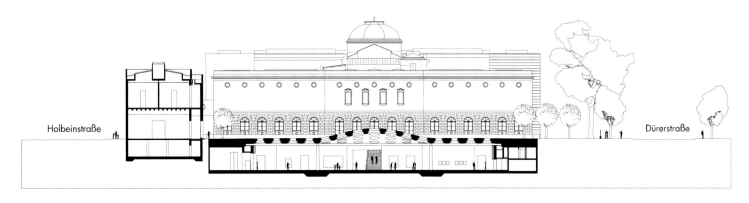

Holbeinstraße   Westflügel   Gartenhallen   Dürerstraße

0 2 5 10 20

Section 剖面图

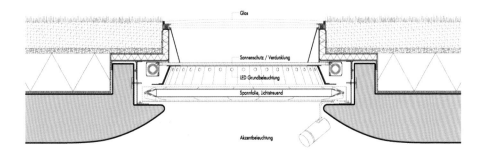

Section Detail 剖面细节图

Design Highlights : Circular Skylights Allow Plenty of Nature Light

The Städel Museum's existing rooms are marked by their abundance of natural light. The new space will be fitted with 195 circular skylights varying between 1.5 and 2.5 m in their circumference, lending the space a similar bright, airy feel to the "old" rooms.

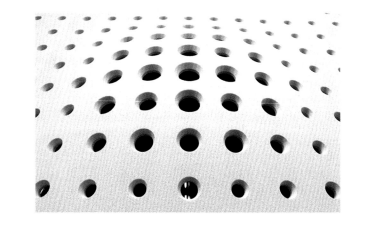

设计亮点：环形天窗引入充沛的光照

新馆室内因其充沛的光照而引人注目，镶嵌在天花上的 195 个直径在 1.5 m~2.5 m 之间的环形天窗，给人以空间明亮、轻纱之感，毫不逊色于"老"房间。

The openings will also include a shading system to avert direct sunlight, while a blackout feature affords the possibility of blocking out daylight completely. The ambient lighting will be integrated into the skylights and individual outlets will guarantee a great deal of flexibility to illuminate individual exhibits. These swellings in the land, bewildering and natural at the same time, will function to enhance the Städel Museum's architectural identity. This green dome alone will constitute a significant enrichment to the architecture of the original Städel Museum complex.

这些天花板开口也包括一个遮阳系统，扭转直射阳光，同时暗视的功能可将日光悉数屏蔽。环境照明系统包含了进顶灯，单体的出口保证了单个展品在照明上极大的灵活性。鼓起的室外地表，迷惑眼球的同时又显得自然，在提升施塔德尔博物馆的建筑个性上功莫大焉。单就一个绿色穹顶，就丰富了原有博物馆建筑群。

The architects have succeeded in subtly offsetting the current separation of the building and the garden area and by extending the space's trajectory out into the garden; they have also created something of an extension of the museum's foyer.A path leading through the grounds reveals ideal resting spots, sculptures, areas of retreat and spaces to hold events. The new design of the garden area could also prove beneficial to the architectonic relationship between the Städel Museum and the Städel Art Academy. The Städel Art Academy, which was modernized in the course of the construction project by schneider+schumacher, provides a perfect counterpart to the south-facing façade of the garden wing.

建筑师成功地将现有建筑分隔和花园做了偏移，通过将空间轨迹伸至花园，为博物馆的休息厅创造出些许延伸。一条醒目的小径可把游客引至理想的休息处、雕塑、休憩区，以及举办活动的地方。在施塔德尔博物馆与施塔德尔艺术学院之间的建构关系上，花园区域的新设计带来的有利影响将得到证明。因德国schneider+schumacher建筑师事务所的建设而得以现代化的施塔德尔艺术学院，为花园侧厅的南立面提供了一个绝好的对应物。

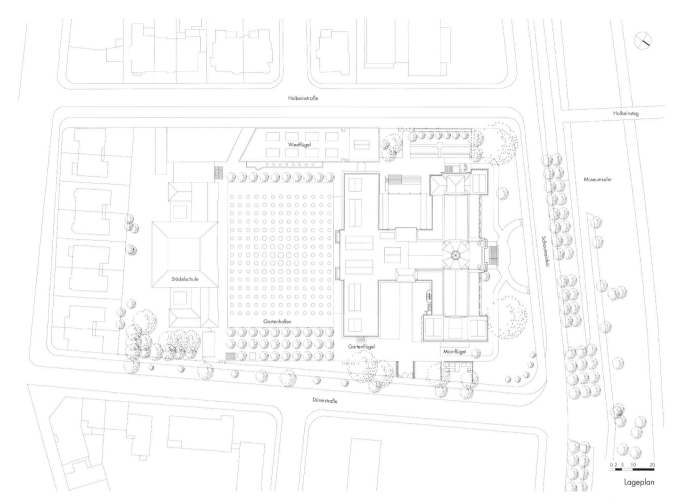

Site plan 总平面图

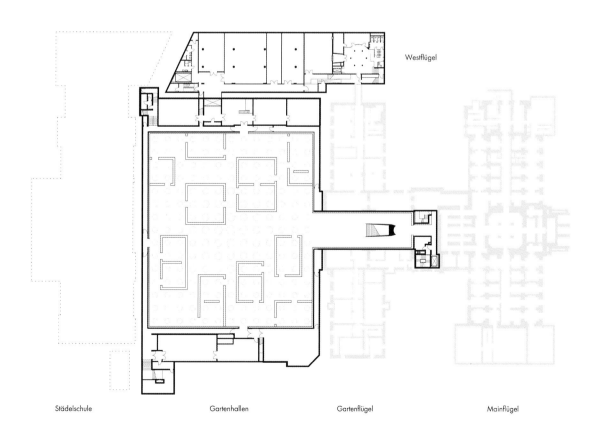

Plan 平面图

In its new form, the garden will span from the building where the art is housed to the courtyard of the building where new art is created. The museum, the Art Academy, the library, the events hall and the garden will form a locus for cultural interaction – conceived as an expression of the progressive mentality of its benefactors.

在新形式中,该花园超越了把艺术收藏在室内的建筑模式,以新的庭院模式绽放艺术魅力。博物馆、艺术学院、图书馆、活动厅和花园形成了文化互动的交点,为奉献者前卫的设计理念创造了完美的表达。

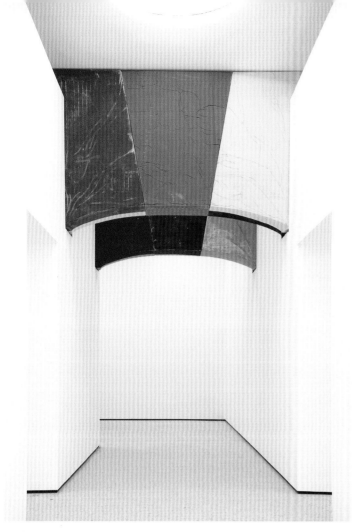

◯ Design Details

The garden halls' interior then will be characterized by the elegantly curved, seemingly weightless ceiling, spanning the entire exhibition space.

◯ 设计细节

看上去无重力并且优雅弯曲的天花板横跨整个展览空间，使新馆充满个性。

195 circular skylights varying between 1.5 and 2.5 m in circumference will flood the space below with natural light as well as form a captivating pattern in the garden area above. Outside, the green, dome-like protrusions, which visitors will be able to walk across, will lend the Städel garden a unique look and create a new architectural hallmark for the museum. "Frankfurt will not only gain a new, unique exhibition building", declared the competition jury, "but as a 'green building' it will also be very much abreast of its times." The generously spacious, light-flooded garden halls will be the new home of the contemporary art section of the museum's collection.

天花板上镶嵌了195个直径在1.5 m~2.5 m之间的环形天窗，使地下展览空间总是光线充足，也使得地上花园展现出迷人的地貌。俯瞰花园，映入我们眼帘的是那略微突起的、能承载人的绿色穹顶，它不仅赋予施塔德尔花园独特的外观，也为博物馆创造出新的建筑标志。竞选评委是这样评价新馆的："法兰克福不仅得到一座全新而独特的展览建筑，而且也得到了一座非常贴合于时代的'绿色建筑'。"充裕而宽阔的、充满阳光的新馆将成为博物馆当代艺术展区的新家园。

Designed by Oskar Sommer in 1878, the interior of the museum's original historical building, the Main River wing, on Schaumainkai is organized around a central axis. A second construction phase to add the two garden wings in 1921 served to extend this axis in keeping with Sommer's original concept. In light of this long reaching history it seems only natural to maintain this established principle and extend the trajectory along the central axis via the Metzler Foyer and into the new exhibition space.

施塔德尔博物馆最初是在 1878 年由奥斯卡·索默设计的，原有历史建筑的内部、美因河侧厅沿 Schaumainkai 街围绕着一条中轴线组织起来。第二期建筑在 1921 年加建了两个花园侧厅，在延长轴线的同时仍保持索默的原概念。这段悠长的历史对博物馆有着较深的影响，在此背景下，只有维持该既有原则才显得自然。

The extension will ascribe the entrance hall in the original building and its central staircase a particular significance. By opening the two tympanums to the right and left of the museum's main entrance foyer, visitors will be able to reach the Metzler Foyer level in the garden wing. Together with the adjacent Metzler Hall, currently holding Thomas Demand's new installation, the Metzler Foyer will also function as an event venue as well as providing additional exhibition space.

最好的方法是将该轨迹沿中轴线延伸到梅茨勒厅，引入新展览空间。在扩建工程中，施工人员将门厅归入原有建筑，并赋予其中央楼梯的特殊意义。山墙弧面开向博物馆主门厅的左右两侧，将参观者引达花园侧厅内的梅茨勒休息厅。这使得梅茨勒休息厅成为举办活动的场地和额外的展览空间。目前加建的展览主体空间与梅茨勒展厅一起，展示托马斯戴芒德的最新装置作品。

An underground energy storage unit based on heat exchangers (geothermic drilling down to 90 m) and downstream heat pumps will be used to heat and cool the museum's rooms as necessary. This energy storage unit will help balance out seasonal variations in the museum's energy requirements, while the heat pump will subsequently allow the museum to cover its heating and a portion of its cooling requirements using renewable energy sources. The planned ventilation system will not only cool the newly-built exhibition halls but humidify and de-humidify them too. It is also equipped with a highly efficient heat recovery facility, while diffuser outlets in the walls will aerate the space. The technical components will be housed in a control room adjacent to the exhibition halls. The compact underground construction, the energy storage unit for heating and cooling the space and the large internal storage capacity will create the optimum internal climate for such a building while using as little energy as possible.

热量交换器（下钻90 m的地热）的地下能量储备，平衡了因季节变化而变化的博物馆能量需求，同时热泵通过可再生能源，使博物馆达到全馆制暖及部分地方制冷的需求。已规划好的通风系统不仅能让新建展厅降温，也能为之增湿和除湿。同时配备高效热恢复设备，墙体里的散气口也可以向展厅充入空气。技术配件布置在展厅相邻的控制室内，密集的地下构筑、供热致冷用的能量储备单元与大型内部储藏室一起，在保证最小能耗的同时，取得最大的优化效果。

- **Architects:** *Claudio Lucchin /Cleaa*
- **Main Contractor Group Leader:** *ZH GENERAL CONSTRUCTION COMPANY AG*
- **Photographer:** *Alessandra Chemollo*
- **Location:** *Bolzano, Italy*
- **Area:** *2,030 m²*

- 设计公司：*Claudio Lucchin /Cleaa*
- 主要承建团队负责人：*ZH GENERAL CONSTRUCTION COMPANY AG*
- 摄影师：*Alessandra Chemollo*
- 地点：意大利博尔扎诺
- 总面积：2 030 m²

汉娜·阿伦特职业学校地下学校扩建

"Hannah Arendt" Professional High School Underground Extension

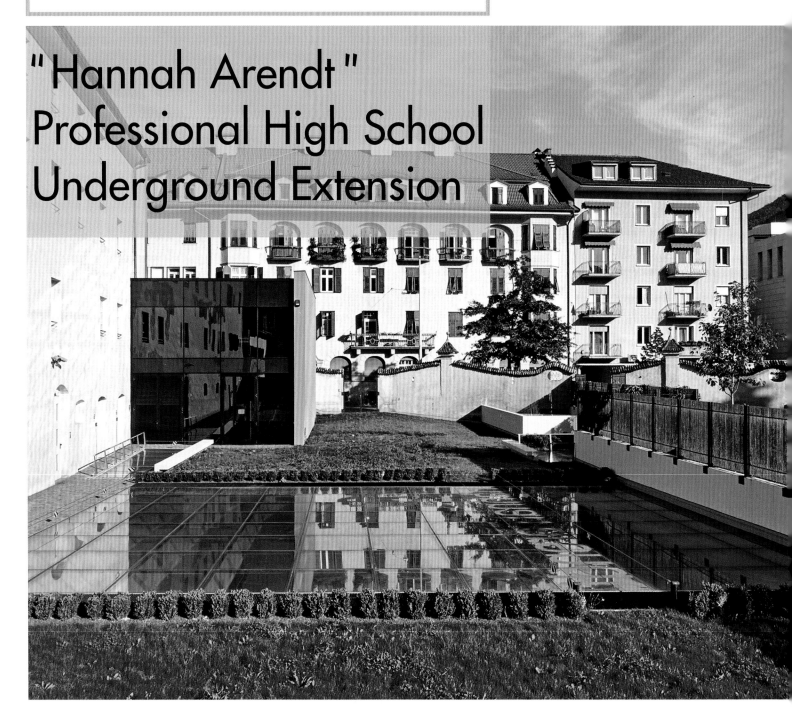

◯ Project Overview

Hanna Arendt school in Bolzano is the first underground school in Italy.

Designed by Cleaa Claudio Lucchin & architetti associati as the extension of the professional existing school, it highlights the unexpected potentialities of the underground architecture, challenging the limits of the sustainability culture as we thought so far, as well as the contemporary design in historic centres.

◯ 项目概况

这是位于博尔扎诺的汉娜·阿伦特学校是意大利第一所地下的学校。Cleaa Claudio Lucchin & architetti associati 为这所职业学校设计了扩建工程。设计师着重挖掘地下建筑的潜力，挑战可持续性文化建筑的极限和传统理念，作出了在历史中心区建造现代建筑的大胆尝试。

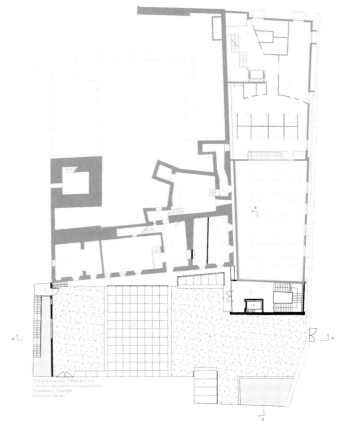

Ground Floor Plan 一层平面图

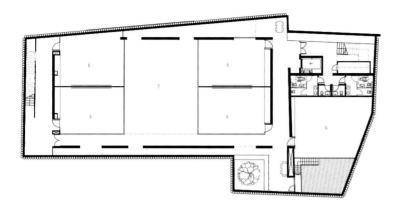

Basement First Floor Plan 地下一层平面图

The idea of not altering the ancient architectural context of the Capuchin friars convent-protected by the national heritage association- but the need for new spaces and classrooms, gave the architect the opportunity to create a "subterranean school appendix".

设计要求在不改变嘉布遣会女修道院（受国家遗产协会保护）原貌的前提下建构新的空间和教室，由此设计师设计了一个"地下附属学校"。

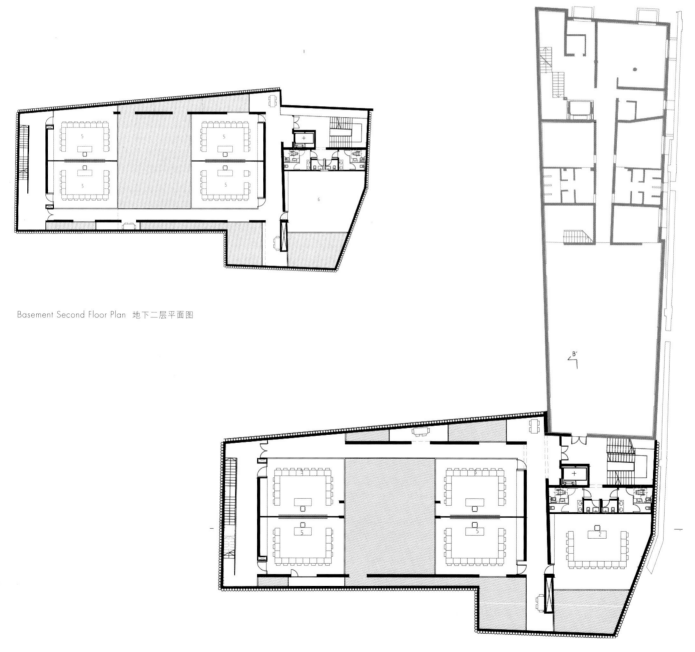

Basement Second Floor Plan 地下二层平面图

Basement Third Floor Plan 地下三层平面图

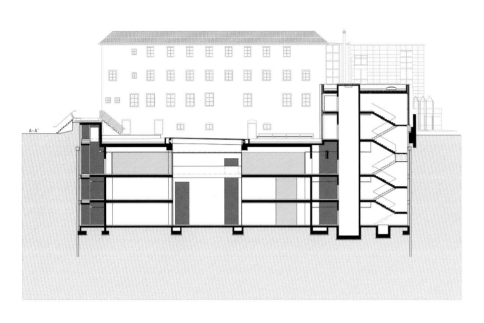

Section 剖面图

Four levels excavated 17 m underground in which 9 classrooms, 6 workshops, a winter garden and a utility room are placed. A big challenge that led the architects to solve consequent problems not only as structural, but particularly enviromental issues.

他们往地下开挖，建造17m深的4层高建筑，这座建筑有9间教室，6个工作室，1个冬季花园和1个多功能室。建筑师面临着一系列巨大的挑战，他们不仅要解决随之而来的结构问题，而且要解决环境方面的问题。

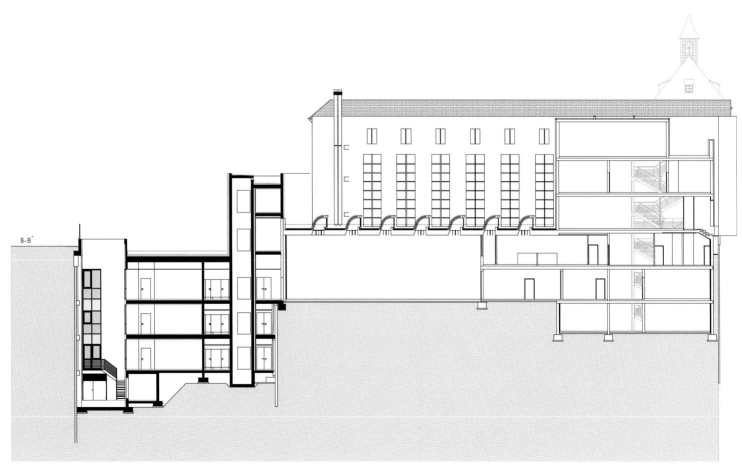

Section 剖面图

● Design Highlights : The Intervention of Lighting

Thanks to the glass walls, many viewpoints allow to perceive the building in all its depth. The central court and the full-height voids play with the materiality of the exposed concrete walls; the connection elements, such as the big yellow staircase and the walkways punctuate the whole space; alongside the paths, numerous niches have been created as small private rooms for studying.

● 设计亮点：采光介入的空间

由于该建筑采用了玻璃墙，人们的视线能达到内部的各个层面。中央庭院和所有楼层的天井与裸露的混凝土墙虚实相生；一些连接的元素如大的黄色楼梯和过道，令整个空间变得更醒目；沿着过道，排布着各种壁龛式的空间，都被用作小的个人学习室。

The lighting design was one of the main topic of the intervention : a costant use of glazed surfaces, in the large skylights and glass walls of the rooms, lets natural light penetrating all internal spaces, allowing a special, continuous visual connection with the outside.

项目的采光的设计主要采用介入的方式：不断的利用屋顶的大玻璃天窗，房间的玻璃墙，让自然光线充满室内空间，也和室外空间产生了一种特殊的视觉上的连续。

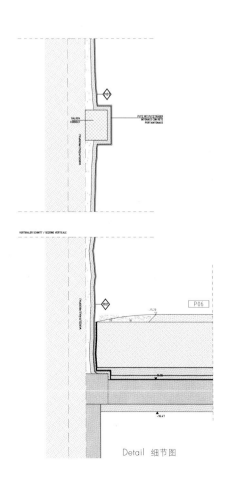

Detail 细节图

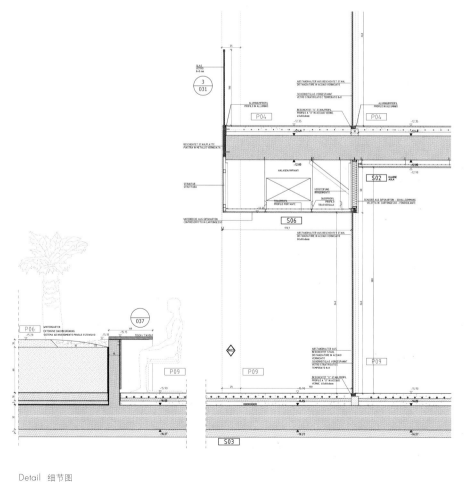

Detail 细节图

Therefore the atrium garden, the small winter garden and a series of skylights and solar chimneys give more light and air to the whole building. The artificial lighting is controlled by sensors neon varying temperature based on time of day and the weather conditions outside.

另外，天井花园，也就是一个小小的冬季花园，以及顶部一排天窗和太阳能烟囱也让整个空间变得更加明亮开敞。人工照明系统由氖气传感器控制，随温度变化而变化，每天不同的时间和外面不同的天气状况，对其有着不同程度的影响。

○ **Design Details**

Humidity has been removed inserting in the walls several layers consisting of insulation, sheathing and plaster spray that also provides protection against ingress of radon gas; to recall the excavation the walls have an irregular surface.

○ **设计细节**

室内的墙面由绝缘层、夹衬板和石灰抹面构成，能有效除湿，并防止氡气的进入；墙面是凹凸不平的，是因为为了更好地除湿，墙面经过了开凿、修建。

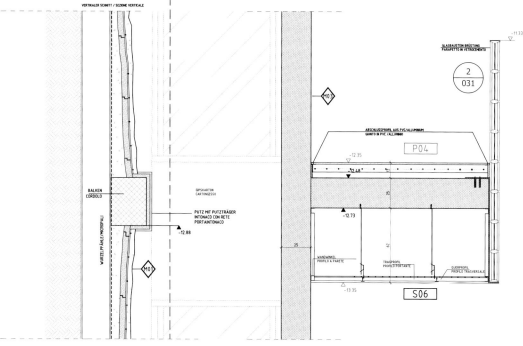

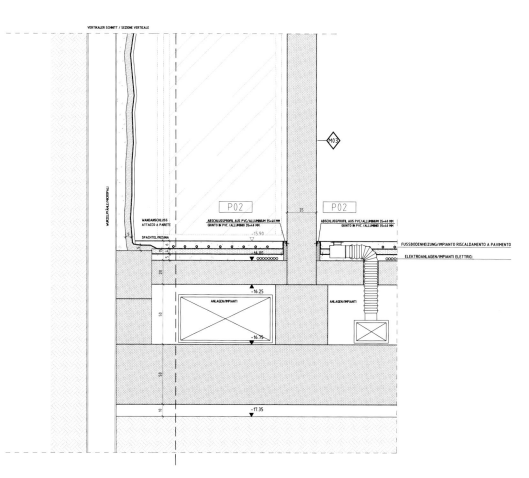

Plan| 平面图

Ventilation is guaranteed by programmed recycling of the air regulated by a mechanical system through ceiling diffusers or grilles integrated into built in wardrobes.

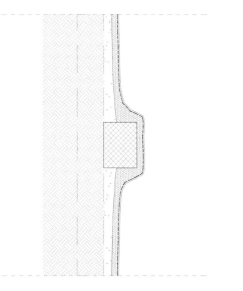

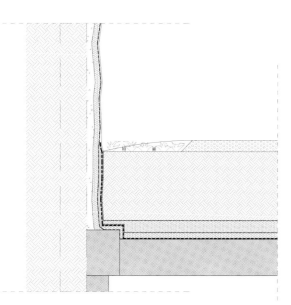 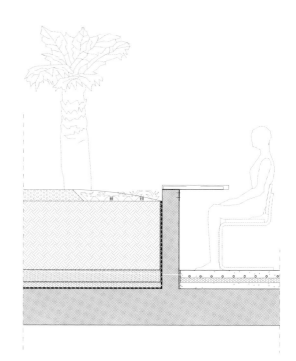

Detail 细节图

由天花板扩散器和衣柜的网格装置构成的机械系统保证了空气的循环流通。

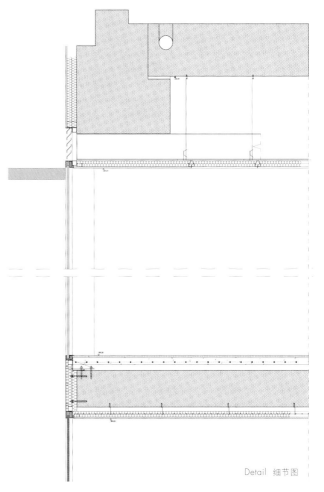

Detail 细节图

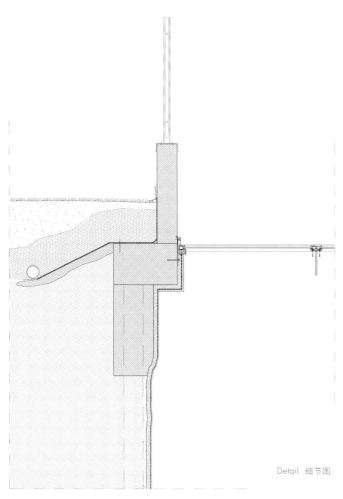

The connection between the old building and the new one takes form of an extension to the existing corridor located at the first floor. Lit through large glasses, and enclosed by a wall acting as a scenic backdrop, this extension features as the only new architectonic element visible above ground.

新老建筑之间的连接，在形式上和位于一层的现有的走廊结合起来。因此在一层只能看到新建筑的玻璃顶和底下用墙围合的空间。

Detail 细节图

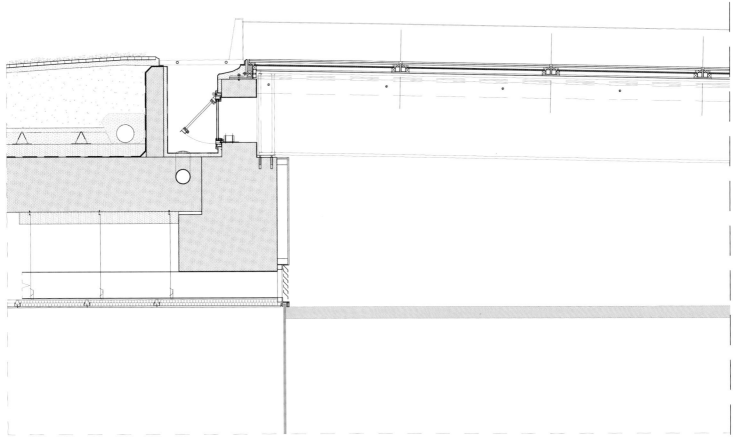

Detail 细节图

The four underground floors were built after an initial stabilization of the area with micro poles and a reinforced concrete structure. The rooms are distributed around the central void; starting from the top the first 2 floors host classrooms; the third floor hosts the workshops and the last one is an utility room.

地下的四个楼层,是建立在用微杆、混凝土搭建起来的地方完成初步加固的基础上的。房间是围绕中央的天井分布的:从顶层算起,第一、二层为教室,第三层是工作室,最底下一层为多功能室。

- **Architects:** *Spora Architects*
- **Design Team:** *Tibor Dékány, Sándor Finta, Ádám Hatvani, Orsolya Vadász, Zsuzsa Balogh, AttilaKorompay, Bence Várhidi, Noémi Soltész, András Jánosi, Diána Molnár, Károly Stefkó*
- **Client:** *Budapest Transport Ltd., DBR Metro Project Directory*
- **Photographer:** *Tamás Bujnovszky, Tibor Dékány, Adam Hatvani, Sergio Garcia*
- **Location:** *Budapest, Hungary*
- **Gross Area:** 7,100m²
- **Site Area:** 3,000m²

- 设计公司：*Spora Architects*
- 设计团队：*Tibor Dékány、Sándor Finta、Ádám Hatvani、Orsolya Vadász、Zsuzsa Balogh、AttilaKorompay、Bence Várhidi、Noémi Soltész、András Jánosi、Diána Molnár、Károly Stefkó*
- 客户：*Budapest Transport Ltd.、DBR Metro Project Directory*
- 摄影师：*Tamás Bujnovszky、Tibor Dékány、Adam Hatvani、Sergio Garcia*
- 地点：匈牙利布达佩斯
- 总面积：7 100m²
- 占地面积：3 000m²

布达佩斯 M4 Fővám tér 地铁站

M4 Fővám tér - Underground Sation, Budapest

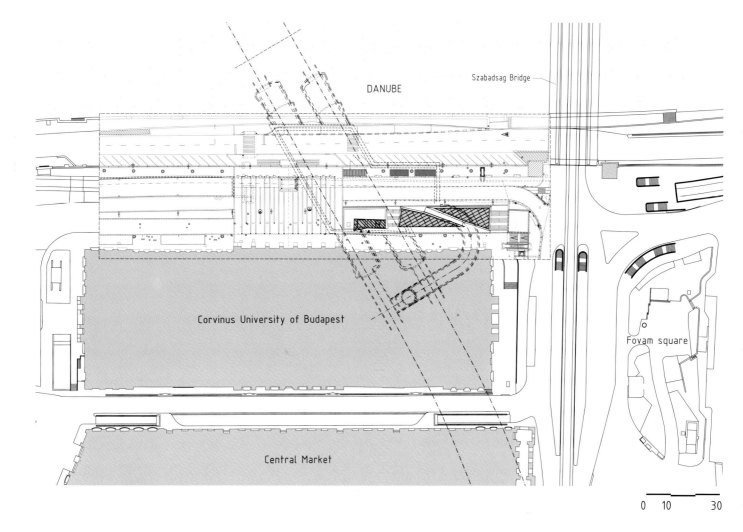

Site Plan 总平面图

◐ Project Overview

The new metro line planned in Budapest is to connect South-Buda with the city center. 10 stations is constructed in the first step, the line is 734 km long. Fővám tér station is one of the deepest stations of the line; it is situated below the bank of river Danube on the Pest side.

◐ 项目概况

计划建在布达佩斯的新地铁线路连通南布达和城市中心。在第一期工程中,10个站点已经建好,这条线路全长734 km。Fővám tér 地铁站是这条线路最深的地铁站之一。它位于城市佩斯的一侧,在多瑙河河岸的下面。

Site Plan 区位图

Budapest, Fővám tér metro station

Budapest, M4 metro line

The station is in a very special situation, because of its place in the city and because of the Duna River. The depth of the platform level is 36 m from the surface. The complexity of the structure is even greater, since here a new tunnel for the tramline and a new pedestrian subway has to be constructed. The Fővám tér station is more than a simple metro station, it is a complex traffic junction of the city, an interchange spot for tramways, buses, metro, ships, cars and pedestrians, which altogether create a unique spacey public space above and under the ground. Fővám tér metro station is new multilevel city junction, gateway of the historic downtown of Budapest.

地铁站的情况非常特殊，这和它所处在这座城市的地理位置有关，也和多瑙河的历史文化背景有关。站台离地面36 m。自从这里建成了一个电车隧道和地下人行通道以后，建造的难度加大了。Fővám tér地铁站不是一个简单的地铁站，它是一个复杂的城市交通枢纽，也是一个为有轨电车、公共汽车、地铁、轮船、小汽车和行人而设的换乘点，这些因素使得这个地方，无论是地上，还是地下，都变成了一个特别的、错综复杂的公共区域。Fővám tér地铁站是一个新型的、多层次的城市交界点，也是通往历史悠久的布达佩斯市中心的门户。

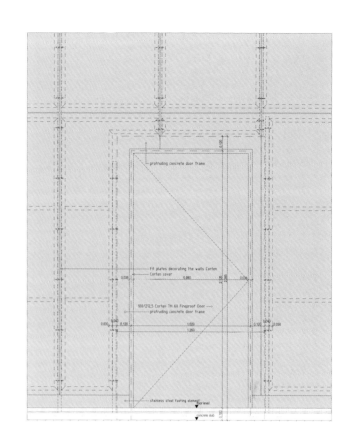

Detail 细节图

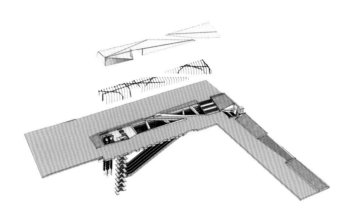

Design Concept 设计概念图

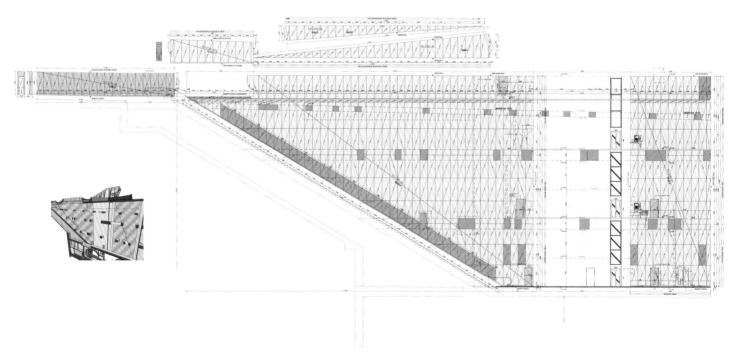

Section 剖面图

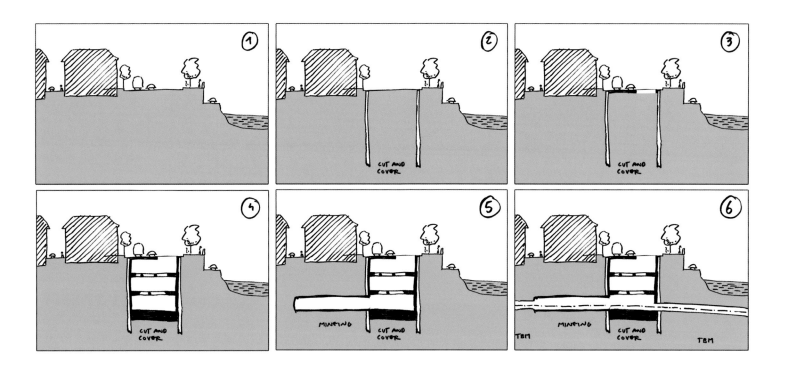
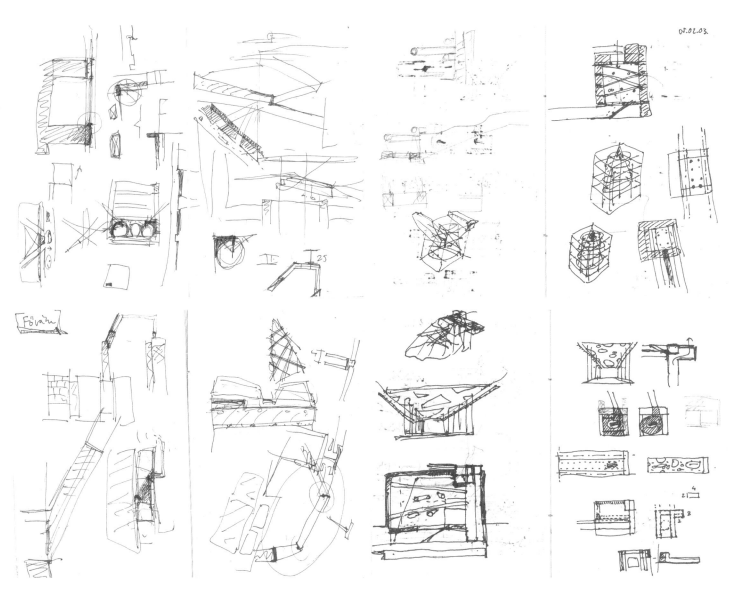

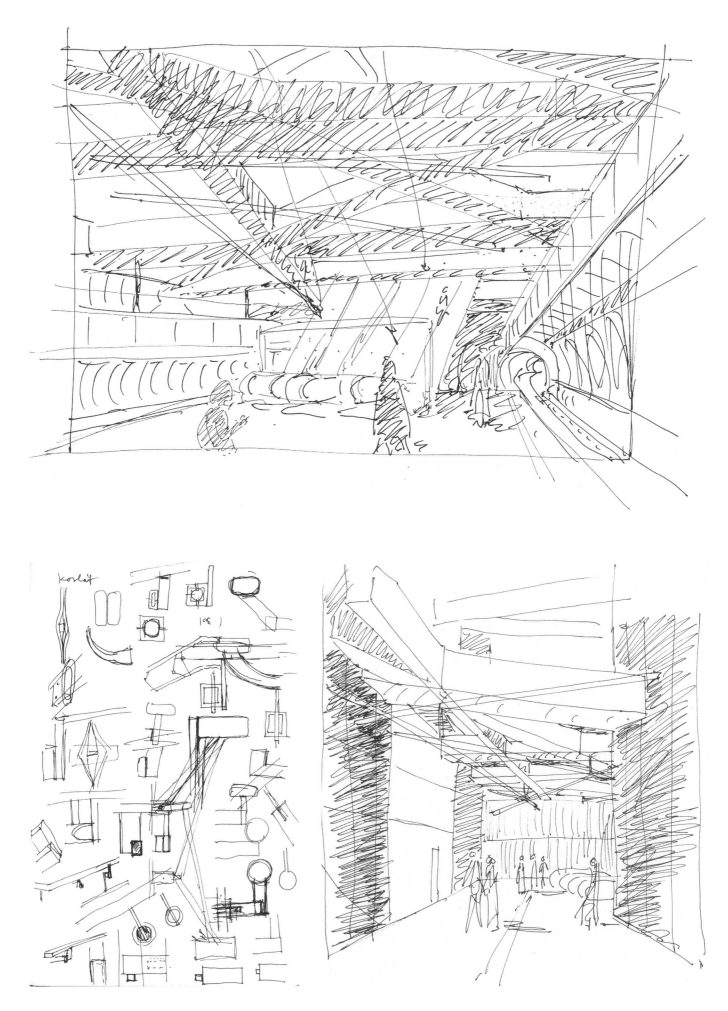

Lighting Schematic Diagram 采光示意图

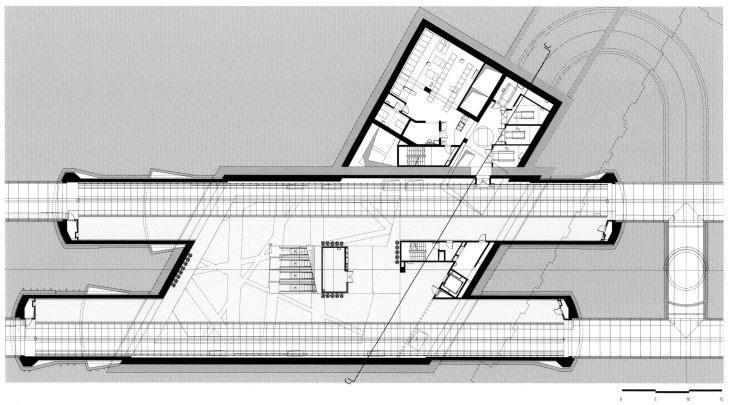

Plan 平面图

◯ **Design Highlights :Collecting Light by the Underground Box**

The main character of the stations is defined by the cut-and-cover technology, resulting in large underground box for the stations. The common architectural decision was to clear this box keeping only the necessary rough structures in the headspace of the stations. This huge reinforced concrete beams can be lit by natural light from above, if possible, providing an extraordinary effect in the deep stations.

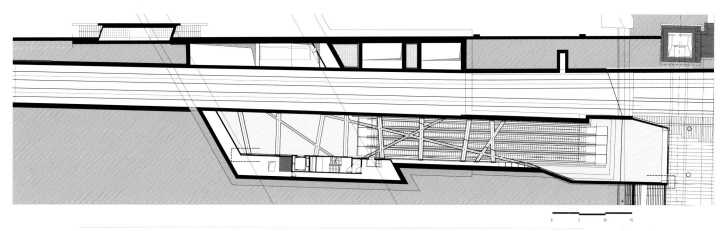

Plan 平面图

◯ 设计亮点：盒状结构采光

这个地铁站的主要特征是采用了明挖回填技术，其结果是让车站有一个盒子状结构。整个建筑设计明确了盒状结构的作用，仅仅保留了车站顶部空间一些必要的建筑毛坯。因此，从上面照射下来的自然光线能够照射到用钢筋混凝土搭建起来的梁柱，有条件的话，还能为位于地下深处的车站带来不同寻常的效果。

Due to the construction technology, huge rooms have been created in the inner spaces of the stations. The section of the space is proportional to cross section of average street in downtown Pest, built in the eclectic period in the 19th century, so the station can be interpreted as an inverse street or square under the surface.

出于对施工技术的考虑,在地铁站的内部已经预留了巨大的空间。地铁内部空间按一定比例与在佩斯市中心的、建立在兼收并蓄的19世纪的普通街道交叉连通,因此该地铁站会被视为一条逆向的街道或一个地下的广场。

Playing on natural light has been an important aspect in the architectural formation of the entire line. On the surface of Fővám tér a huge square will be created without traffic. This allows of locating glassy, crystalloid skylights, which let the sunlight reach the platform level, emphasizing the unique character of the beam network.

自然光照是整条线路需要考虑的重点。车站上方巨大的广场是不通行车辆的，这样天窗渗透的光线就能够引入到站台上，并映射出梁柱结构的独特风格。

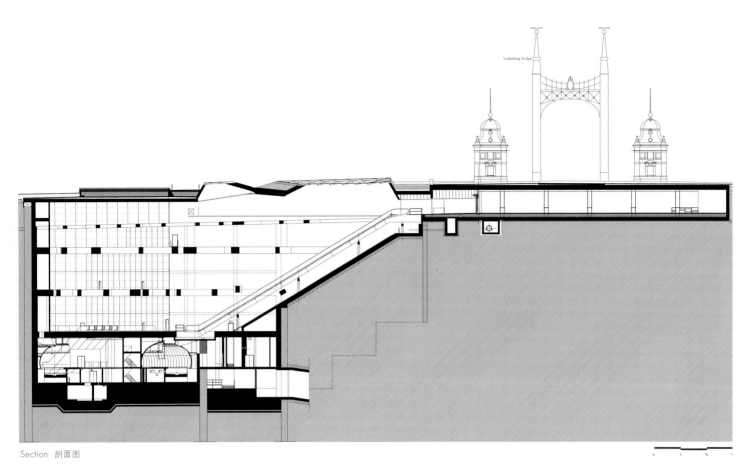

Section 剖面图

○ **Design Details**

The Fővám tér station is composed of a cut-and-cover box and tunnels. Only a complex structural system could fulfil the requirements emerging from the proximity of the Danube and the given construction site.

○ 设计细节

Fővám tér 地铁站由以明挖回填法建造起来的盒状建筑和隧道组成的。只有一个复杂的结构体系能满足连接多瑙河邻近地区和指定的施工现场的要求。

The box is supported by 4 levels of reinforced concrete beams, which structure is made like a net. There are three layers of this network; they keep the walls of the box like a bone-structure. The design of the box is determined by this visible concrete net-structure. In the main front of the box, which is a concrete wall covered with corten steel, runs the elevator with glass walls faced to the inside, in order to connect both visually and physically the parts of the building with the surface.

盒状建筑是由4层的由钢筋混凝土搭建的梁柱支撑起来的，这是一个网状结构。这个有3个层次的网状结构使得盒子部分的墙面如同骨架。地铁站的结构设计由这条线路的纵向的道路定线所决定的。在盒状建筑前面的主要区域，也就是在那个有一堵由混凝土建成的、上面铺着低合金高强度钢块的墙壁的地方，设计师让带有玻璃墙的电梯朝向内部运行，以便让地铁内部在视觉上和工程上更好地连接外面。

In the other part the tunnels have curved cross section. The walls and the columns will be covered with mosaic tiles artwork reflecting to the Zsolnai ceramic tiles of Gellért hotel, which is nearby the site.

隧道的另一部分是一个弯曲的横跨结构,墙面和梁柱用富有艺术风格的陶瓦覆盖,能很好地映衬场地附近 gellért 饭店的陶瓷艺术。

The architectural and structural concept of Fővám tér based on the indiscriminate beam grid and the underground bonc texture combined with the organically implemented construction system were compatible without compromises with the often volatile and changing conditions of the planning and building processes. It was proved to be adjustable to all the emerging technical problems without having lost from its original force at all.

Fővám tér 地铁站的结构设计理念是以不规则的、呈网状结构的梁柱和地下的、呈骨架状结构的墙面为基础的。其中，骨架状的墙面的建造结合了机制较为完善的、实施情况良好的施工系统。这样的施工系统兼容性强，且不会因为规划和建筑过程中所出现的经常不稳定、经常改变的条件而妥协。很多新出现的技术问题可以在不影响原有条件的前提下进行调整。这一点已经得到证实。

It's been highly inspiring to create spaces to be used with pleasure, so that the passengers might prefer the public transport somehow more. It's important to emphasize that it's supposed to be a public space – a public space under the ground. Also for that reason, the public activities are wellcome in the spaces of the stations, even during the time of the construction. People can use the space for art and new media related projects.

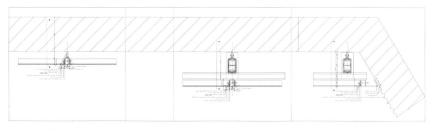

Detail 细节图

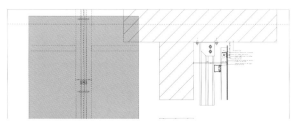

Detail 细节图

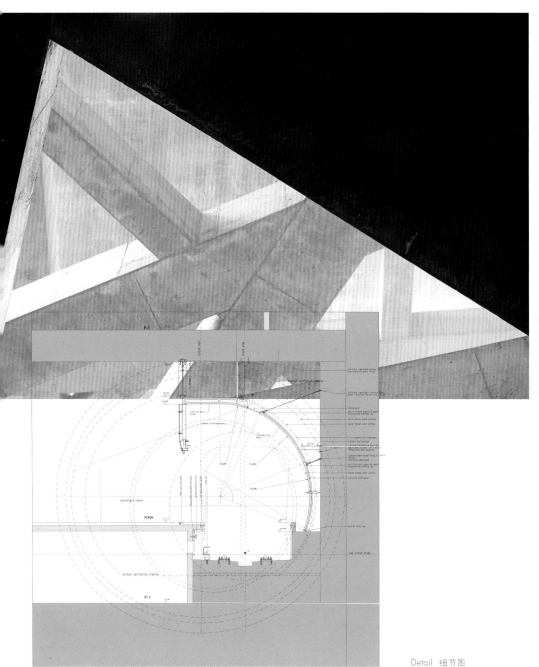

Detail 细节图

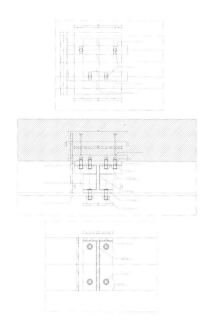

Detail 细节图

Detail 细节图

创造令人愉悦的空间，让乘客们更喜欢以某种独特形式呈现的公共运输系统，这真是一件鼓舞人心的事情。更重要的是，有一件事情需要强调，那就是这本应该是个公共空间———个位于地下的公共空间。也因为这个原因，公共活动都可以在地铁站的这些空间举行，哪怕在施工时间也行。人们可以利用这些空间进行艺术活动和与媒体相关的活动。

- **Architects:** SOLDEVILA, SOLDEVILA, SOLDEVILA ARQUITECTOS S.L.P
  Alfons Soldevila Barbosa,
  Alfons Soldevila Riera y David Soldevila Riera, arquitectos
- **Building Company:** UTE-ARQUITECTURA GORG (DRAGADOS, ACCIONA, COMSA, ACSA)
- **Photographer:** M. Grassi
- **Location:** Barcelona, Spain

- 设计者：SOLDEVILA, SOLDEVILA, SOLDEVILA 建筑事务所
  Alfons Soldevila Barbosa,
  Alfons Soldevila Riera y David Soldevila Riera, arquitectos
- 建造公司：UTE-ARQUITECTURA GORG (DRAGADOS, ACCIONA, COMSA, ACSA)
- 摄影师：M. Grassi
- 地点：巴塞罗那西班牙

La Salut
地铁站

La Salut Metro Station

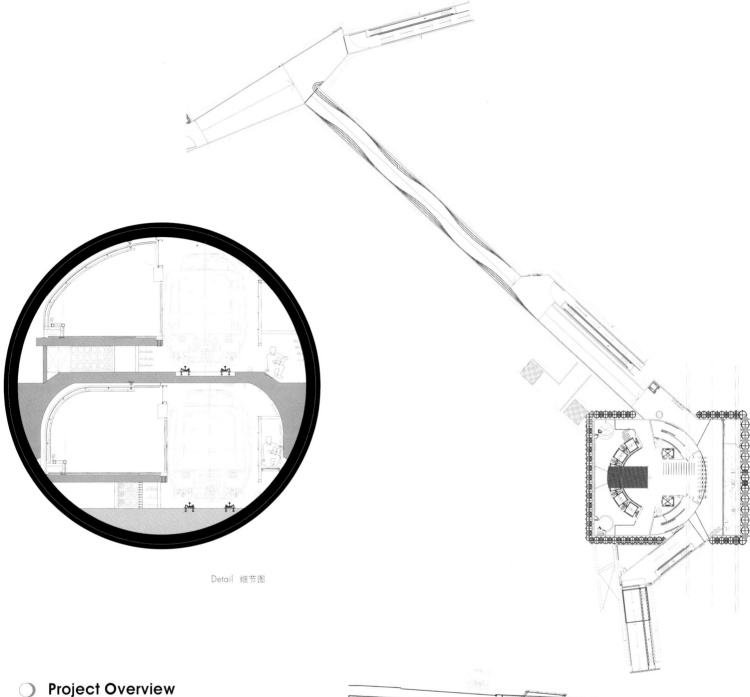

Detail 细节图

◌ Project Overview

The La Salut Station, located in the key point of the city, is one of most important metro stations of the new metro line—line 9 of the Barcelona metro. It plays an important role in the transportation.

◌ 项目概况

位于城市核心位置的 La Salut 地铁站，是巴塞罗那最新建起的地铁线路——地铁 9 号线中的最重要的站点之一。它在交通运输中发挥着重要的作用。

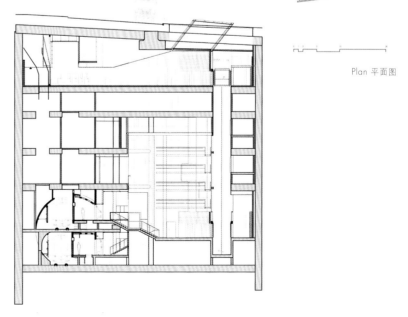

Plan 平面图

Section 剖面图

Design Highlights : Collecting Lights by a Great Skylight

The hall of La Salut station receives natural light through a great skylight, an inheritance of the entrance route of materials during its construction. The entrances to the station connect through hallways to a great artificially illuminated atrium, which then opens to the space illuminated by the skylight. One of the hallways, great in length, uses the inclination of its walls as a mechanism to reduce the monotony of its pass. Each is lined with inlayed galvanized panels.

设计亮点：通过巨大的天窗采光

La Salut 地铁站通过巨大的天窗来采光，根据入口路线情况选用的材料也对采光具有引导作用。从地铁站的入口进入，经过通道，可以来到一个很大的、采用人工照明系统的中庭。从这个中庭可以来到一个通过天窗来采光的空间。其中一条很宽的通道，通过建造一系列倾斜的墙壁让室内空间看起来不那么单调。每一堵倾斜的墙壁里面都镶嵌了镀锌的面板。

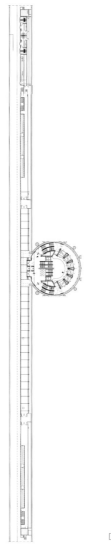

Detail 细节图

◯ Design Details

The station has been solved two essential aspects: the first is reducing the claustrophobia effect and the second is counteracting the vertigo effect in the voids which give access to the platforms. The strategy which is used to minimize these sensations is horizontal fragmentation of space. Claustrophobia is resolved by using back-lit panels on conflictive walls so the space looks breathable. Vertigo sensation is tamed by a sequence of illuminated gangways acting like balconies at a theatre, fragmenting the view of the void. The floor of the gangways is a perforated panel of galvanized steel to create different light qualities.

◯ 设计细节

设计师解决了地铁站两大重要问题：一是减轻了幽闭恐惧症患者的负担，二是减轻了从列车里出来的人的晕车症状。减少乘客可能出现的这些不良感觉的策略是对空间进行水平分割。设计师在这些看起来不协调的墙上安上背光面板，使墙面看起来是透气的，以减轻幽闭恐惧症患者的心理负担。而一系列有良好采光效果的过道的设置，会让乘客感觉到自己就在剧院的阳台上，从而消除不好的感觉。车站的地板是一块用镀锌钢制作的穿孔面板，能营造不同的光感。

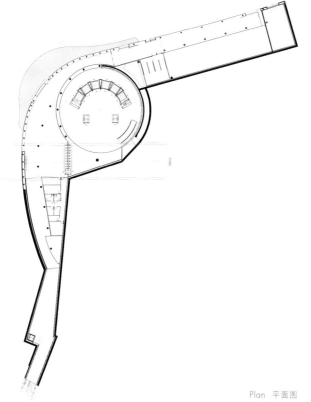

Plan 平面图

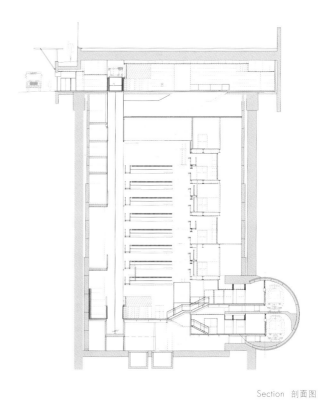

Section 剖面图

- **Architects:** Zechner & Zechner ZTGmbH
- **Client:** ÖBB – Austrian Railways, Vienna
- **Photographer:** Thilo Härdtlein
- **Location:** Vienna, Austria

- 设计公司：Zechner & Zechner ZTGmbH
- 建造公司：ÖBB – 奥地利维也纳铁路公司
- 摄影师：Thilo Härdtlein
- 地点：奥地利维也纳

S7 线
城市地铁站

Urban Railway Stations S7

◯ Project Overview

The urban railway system between Vienna´s centre and the airport got redesigned. Several stations got new configurations and cladding designs. Because Vienna´s International Airport in Schwechat got converted and expanded during the last years, the underground station of the local train line S7, which connects Vienna´s centre with the airport had to be modified in various terms.

◯ 项目概况

设计师重新设计了连接维也纳市中心和机场的铁路系统。一些车站装上了新的配置,其室外部分也进行了重新设计。因为近几年来,位于施韦夏特的维纳斯国际机场进行了改建和扩建,所以这个在 S7 线路上的,也就是连接维也纳市中心和机场的地铁站,需要在各个方面进行改进。

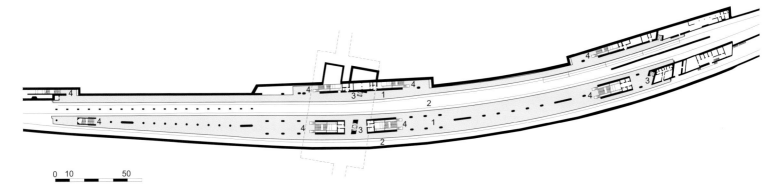

1. Bahnsteig / Platform
2. Gleisbereich offen / Track area open
3. Aufgänge (Aufzüge) / Exit (Lifts)
4. Aufgänge (Treppen, Fahrtreppen) / Exit (Staircases, escalators)

Level Plan 楼层平面图

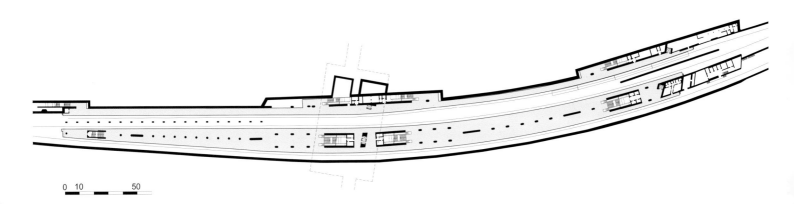

Level Plan 楼层平面图

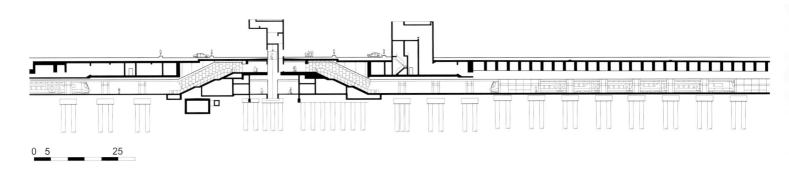

Longitudinal Section 纵向剖面图

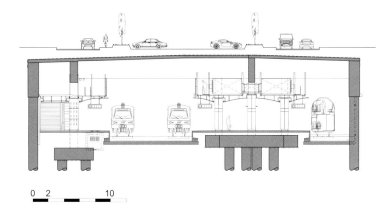
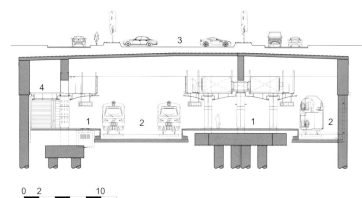

1. Bahnsteig / Platform
2. Aufgänge (Aufzüge) / Exit (Lifts)
3. Parkplatz Flughafen / Parking area Airport
4. Aufgänge (Treppen, Fahrtreppen) / Exit (Staircases, escalators)

Cross Section 横向剖面图 　　　　　　　　　　　Cross Section 横向剖面图

Design Highlights : The Improved Lighting Situation

Much emphasis was placed on improving the lighting situation: A lateral band of light is situated along the platform edge. It results in an optical differentiation of the platform and a better illumination along the platform edge. The faces of the support columns and the entries to the stairs are fitted out with a surface light. The backlight brightens up the space between the supporting structure, improves the lighting situation and enhances the feeling of safety. The orientation is improved by the marking of the staircases.

设计亮点：经改善的照明系统

设计师把更多的精力放在改善照明系统上：沿着站台的边缘有一排横向的灯光带，这将为站台带来不同寻常的光学效果，为站台的边缘带来更好的照明效果。支撑柱的各个面和连接楼梯的各个入口在其表面都安装了灯，背光源照亮了在支撑性结构之间的空间，改善了照明状况，提升了安全感。楼梯间标志的改进，增强了方向感。

○ **Design Details**

The underground station at Vienna's airport got expanded to twice its platform length and was fitted with new exits. These changes provide improved access to and from the airport's new terminal building. There are 2 platforms. One is an island platform, one is a "border"-platform. The central one handles all passengers to the local line, the other one is for intercity rail traffic.

○ 设计细节

设计师已对这个位于维也纳机场的地铁站进行了扩建，它的站台的长度扩展为以前的两倍，与新的出口相匹配。这些调整因机场新的候机楼而生，也改善了机场客流的情况。现在有两个站台，一个是岛型的站台，一个是长板型的站台。中间的那个站台可以引导乘客去往本地的路线，另一个站台则引导乘客乘坐城际交通轨道。

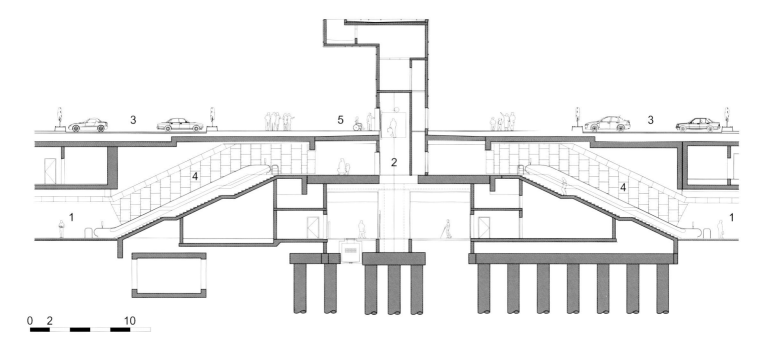

1. Bahnsteig / Platform
2. Aufgänge (Aufzüge) / Exit (Lifts)
3. Parkplatz Flughafen / Parking area Airport
4. Aufgänge (Treppen, Fahrtreppen) / Exit (Staircases, escalators)
5. Vorplatz / Forecourt

Station Exit Section 地铁站出口剖面图

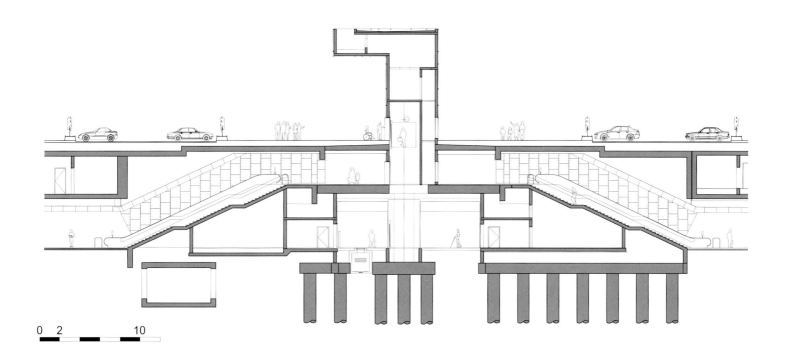

Station Exit Section 地铁站出口剖面图

The new underground station is accessed by three entrances, two of them with escalators and lifts, one of them only with stairs. The configuration allows very quick exits from the station to the airport.

新的地铁站有三个入口，其中两个有自动扶梯和电梯，另外一个只有楼梯。这样的设置让乘客能更快捷地从地铁站去往机场。

The platforms are equipped with a tactile guidance system to enable full disabled access to the station and the airport. The interior design was in keeping with the scheme currently used by the Vienna S-Bahn, which are enameled metal claddings, suspended ceilings in Alucobond and natural stone floors and stairs. The blue colour used by S7-line shows up in several details to provide a good orientation for all passengers.

站台配有容易感知的引导系统，使完全残疾的人能够从地铁站去往机场。室内设计方案保留了维也纳轻轨正在实施的做法，采用金属内墙，并在外面涂上油漆，顶部是用和思瑞安复合材料建造的天花板，地板和楼梯都是用天然石头建造的。S7线在很多细节地方都用上了蓝色，以便为乘客带来更好的导向。

Chapter Two
第 2 章

People have realized that "protection of human memory" is so important in this high technology period. We attach importance to traditional culture, protection of historical relics. On purpose, multi-centers buildings are built for the historic site to coordinate the nature environment. Low visibility of the appearance, provides the advantages of protecting historic architectures and traditional city. It plays a positive role in sustaining the urban cultural vein and reflecting the urban cultural characteristic. The cases in this chapter are base on cultural background. These spaces which are built up for people to use, carry out activities and communicate with each other, reflect the cultural atmosphere of the city.

The Space Based on Culture Protection

人文保护的空间

在这个日新月异的高科技时代，人们已经认识到"保护人类记忆"的重要性，重视文化传统，保护历史遗迹，对具有历史意义的地点注入多中心的用途，并协调自然环境。地下建筑在外观上的低可视性，使它在保护历史性建筑和传统城市风貌方面具有很强的优越性，因此能起到延续城市文脉，体现城市文化特色的积极作用。本章所选取的案例，都是建筑师在充分考虑人们的感情和文化需求的基础上进行设计的。这些供人们使用、活动、交流的主体空间，能体现出城市的文化气息。

- **Design Architects:** C.F. Møller Architects
- **Executive Architects:** Purcell
- **Landscape:** Churchman Landscape Architects
- **Engineering:** Adams Kara Taylor, Fulcrum Consulting
- **Client:** National Maritime Museum
- **Location:** Greenwich, UK
- **Size:** 7,300m²

- 设计公司：C. F. Møller 建筑事务所
- 执行建筑师：Purcell
- 景观设计：Churchman Landscape Architects
- 工程师：Adams Kara Taylor、Fulcrum Consulting
- 客户：National Maritime Museum
- 地点：英国格林威治
- 面积：7 300m²

伦敦国家海事博物馆扩建——萨米·奥弗翼楼

The Sammy Ofer Wing, Extension of the National Maritime Museum, London

◯ Project Overview

C. F. Møller Architects has designed the extension of The National Maritime Museum in London, Britain's seventh largest tourist attraction and part of the Maritime Greenwich World Heritage Site. The new wing, called The Sammy Ofer Wing – named after the international shipping magnate and philanthropist Sammy Ofer.

◯ 项目概况

C. F. Møller 建筑师事务所设计了位于伦敦的英国国家海事博物馆扩建项目，该博物馆为英国第七大旅游胜地，亦是格林威治世界海事遗产的一部分。新翼楼名为萨米·奥弗翼楼，是以国际航运巨头和慈善家萨米·奥弗的名字来命名的。

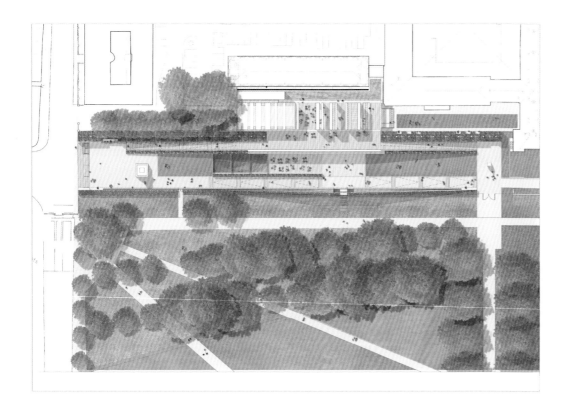

The National Maritime Museum houses the world's largest maritime collection. It attracts two million visitors every year from around the world. The museum is housed in historic buildings, built in 1807, forming part of the Maritime Greenwich World Heritage Site. The park incorporates a number of baroque buildings that are considered among the finest in Europe and is an essential part of Britain's maritime history, particularly The Royal Observatory from 1676 and The Old Royal Naval College from 1712.

英国国家海事博物馆收藏了世界上数量最多的航海文物，每年都有约 200 万名游客从世界各地慕名而来。博物馆坐落在历史悠久的建筑中，这些建筑始建于 1807 年，是格林威治世界海事遗产的一部分。公园里有一些巴洛克式建筑，它们在欧洲首屈一指，是英国航海史的重要组成部分，特别是建于 1676 年的皇家天文台和建于 1712 年的老皇家海军学院。

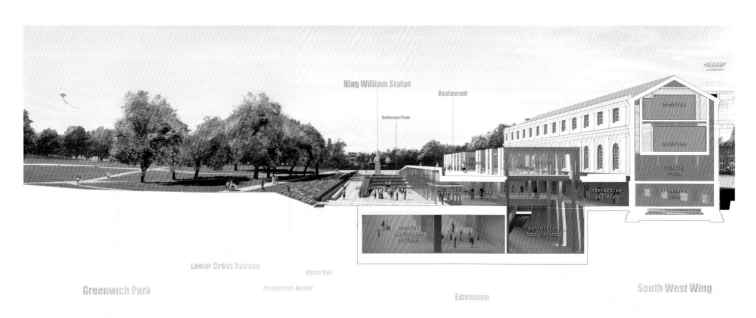

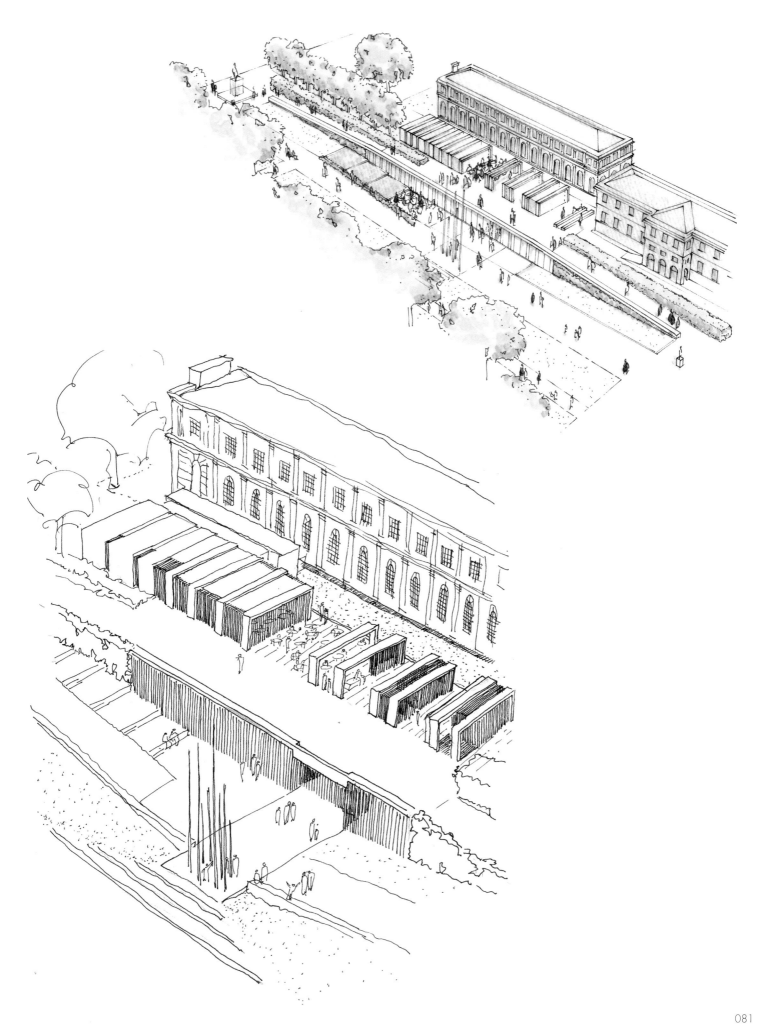

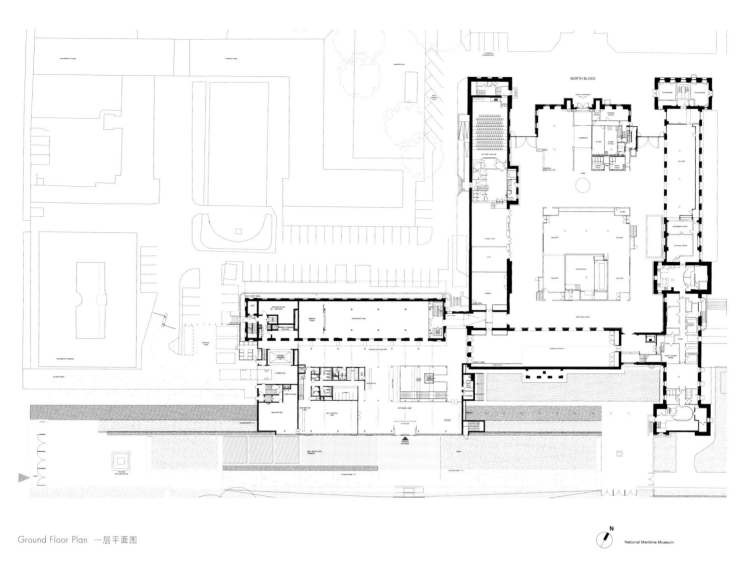

Ground Floor Plan 一层平面图

National Maritime Museum

Design Highlights : The Minimal Intervention

The main idea of the extension - which C. F. Møller Architects won in an international architectural competition in 2006 - has been to ensure minimal interventions in this sensitive historic site and yet give the museum a new, distinctive main entrance and the necessary additional exhibition space, as well as a new café, restaurant, library and archives that meet the particular demands for storage of historical documents. The museum's collections range from e.g. a toy pig that survived the sinking of RMS Titanic to Lord Nelson's last letter to his daughter. The maritime archive contains some 100,000 books and nearly 3.2 km of shelved manuscripts.

设计亮点：最小化介入历史遗迹

该扩建项目是 C. F. Møller 建筑师事务所在 2006 年的国际设计竞赛中的获奖作品，其主要设计思想是确保对这个敏感的历史遗迹形成最小化的介入，同时为博物馆带来全新而又与众不同的主入口和必要的额外展览空间，以及新咖啡馆、餐厅、图书馆，还有能满足特殊历史文件存储需求的档案馆。博物馆的收藏品范围极广，从沉没的泰坦尼克油轮上幸存的玩具猪，到纳尔逊勋爵留给女儿的最后一封信，应有尽有。海事档案包含大约 10 万本书和近 3.2 km 长的暂时搁置的手稿。

The design solution by C. F. Møller Architects has created a new main entrance emerging from the terrain. Due to the heritage sensitivity of the site, the bulk of the new construction is kept below ground, so that new wing becomes a combination of architecture and public landscape. The gross area of the new extension is 7,300 m², and the area of the underground part is 5,500 m². The profile of the new extension has been kept low to allow the Grade I listed Victorian façade of the existing south west wing of the museum to be appreciated as a backdrop to the striking new building.

C. F. Møller 建筑师事务所创造了一个随地形浮现的全新主入口。考虑到建筑所处位置的特殊性，新建筑的大部分地方都位于地下，以便能将建筑和公共景观结合在一起。它的总面积是 7 300 m²，其中地下部分的面积是 5 500 m²。新建筑外形并不高大，以突出博物馆西南翼的一级保护文物——维多利亚时代的外墙，作为新建筑的背景供人欣赏。

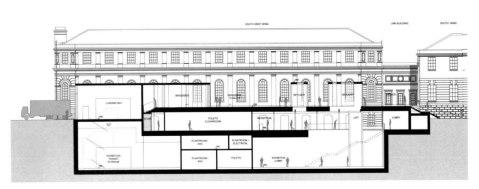

Section 剖面图

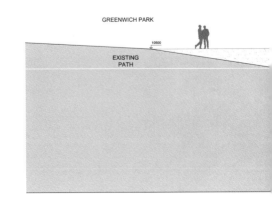

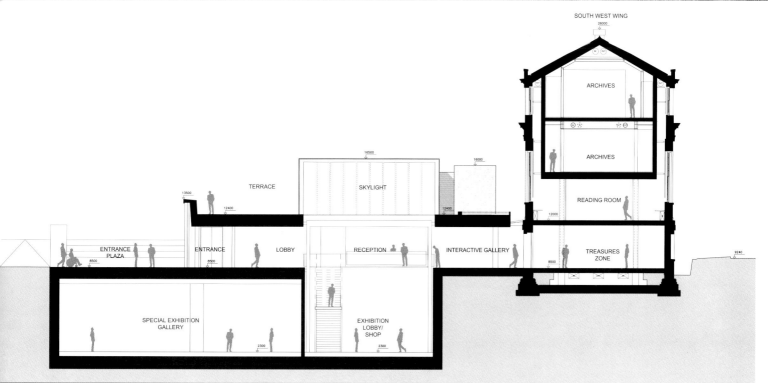

Section 剖面图

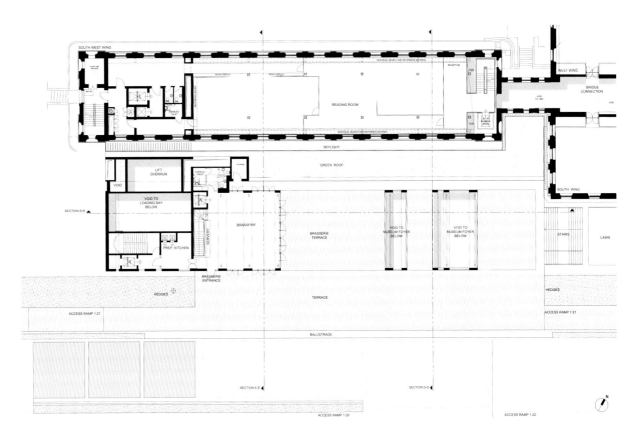

First Floor Plan 二层平面图

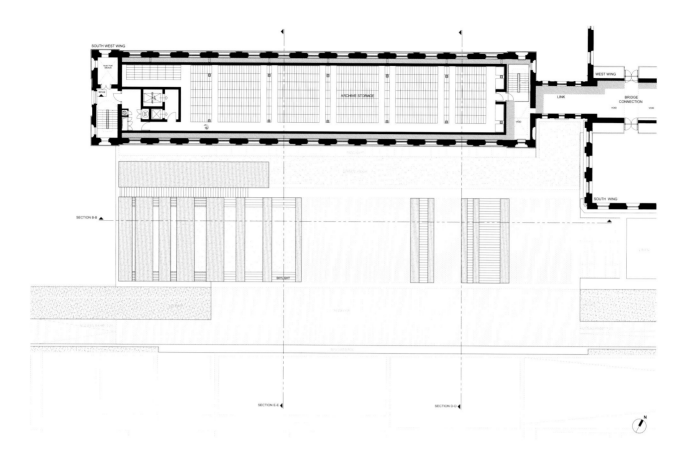

Second Floor Plan 三层平面图

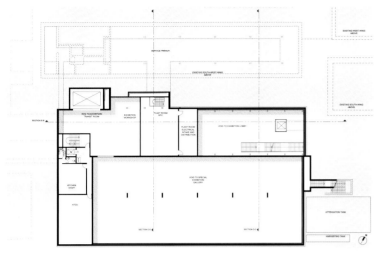

Basement First Floor Plan 地下一层平面图

Design Details

Collaborating with Churchman Landscape Architects, the design team established a concept that looked beyond the immediate building footprint, proposing a new east-west pedestrian route to unify the grounds of the National Maritime Museum with Greenwich Park.

设计细节

设计团队与Churchman景观建筑事务所，根据外面人流走向的情况，提出了一个东西走向的步行路线方案，以便让国家海事博物馆各个楼层与格林威治公园更好地连接起来。

These alterations create a new level of accessibility from the parkland, through the museum's formally landscaped areas, to the new Sammy Ofer Wing.

这些设计上的改变，提高了出行的可达性。游客可以很方便地从公园的场地，穿过博物馆的正式风景区，到达新建的萨米·奥弗翼楼。

- **Architects:** BIG
- **Project Leader:** Ole Schrøder, Ole Elkjær-Larsen
- **Project Team:** Christian Alvarez Gomez Jeppe Ecklon Rune Hansen Thomas Juul-Jensen Narisara Ladawal Schröder Jakob Lange Xu Li Riccardo Mariano Henrick Poulsen Dennis Rasmussen
- **Collaborators:** EKJ, CG Jensen, Jens Linde
- **Client:** Gammel Hellerup Gymnasium
- **Location:** Hellerup, DK
- **Size:** 1,100 m²

- 设计公司：BIG
- 项目负责人：Ole Schrøder, Ole Elkjær-Larsen
- 项目团队：Christian Alvarez Gomez Jeppe Ecklon Rune Hansen Thomas Juul-Jensen Narisara Ladawal Schröder Jakob Lange Xu Li Riccardo Mariano Henrick Poulsen Dennis Rasmussen
- 合作者：EKJ, CG Jensen, Jens Linde
- 客户：Gammel Hellerup Gymnasium
- 地点：丹麦赫勒
- 面积：1 100 m²

Gammel Hellerup 体育馆

Gammel Hellerup Gymnasium

◯ **Project Overview**

Old Hellerup High School, with its characteristic yellow brick buildings, is a good example of a school building in a human scale and a fine architectural example of its time. The sports facilities have, however, become too insufficient and the high school is lacking a large, multifunctional space for physical activities, graduation ceremonies and social gatherings.

◯ 项目概况

古老的 Hellerup 高级中学,以其黄砖建筑为特色。学校有着适合人们学习、活动的建筑规模,也因此成为学校建筑的典范。学校有着悠久的历史,其建筑结构的稳定性也为建筑界所称赞。然而,学校的体育设施非常欠缺,非常需要一个用于体育活动、毕业典礼和社交活动的大型多功能空间。

The Old Hellerup High School, a self-owned governmental institution, wishes therefore to build a new flexible hall for the students' usage with a particular focus on sustainability.

作为一个独立自主的政府机构,古老的 Hellerup 高级中学,希望新建一个以可持续发展为重点的、形式灵活的大厅给学生使用。

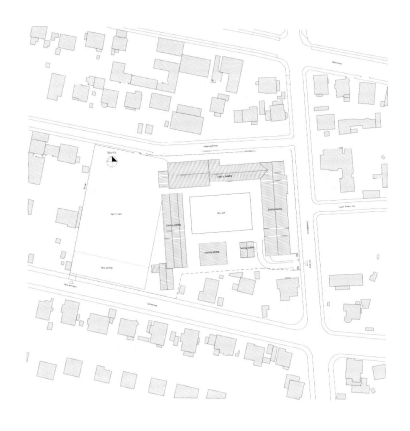

Site Plan 总平面图

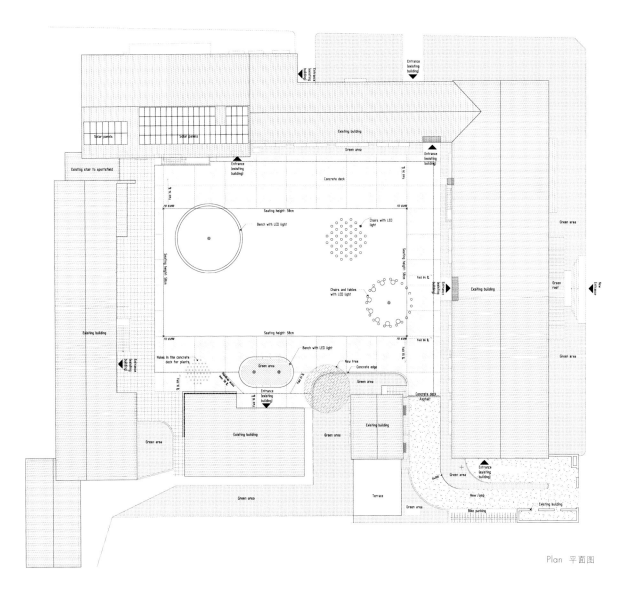

Plan 平面图

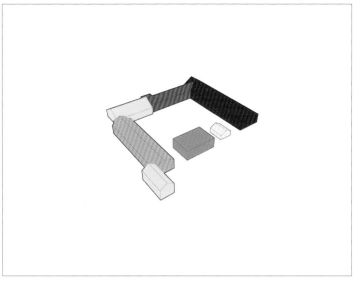

Design Concept 设计概念图

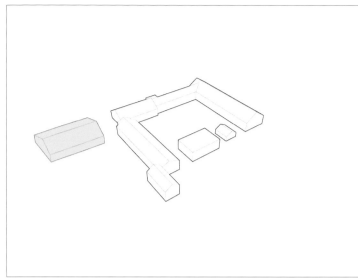

Design Concept 设计概念图

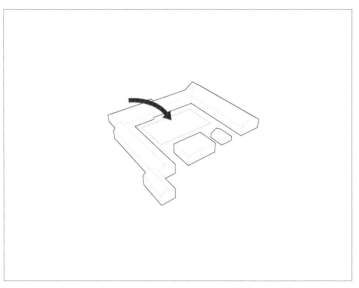

Design Concept 设计概念图

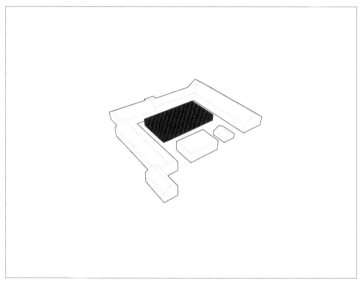

Design Concept 设计概念图

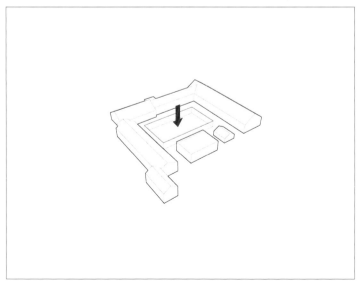

Design Concept 设计概念图

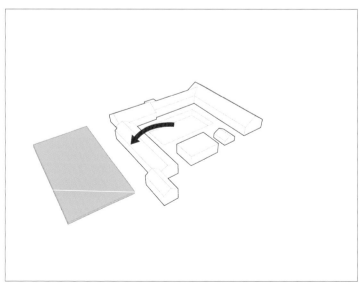

Design Concept 设计概念图

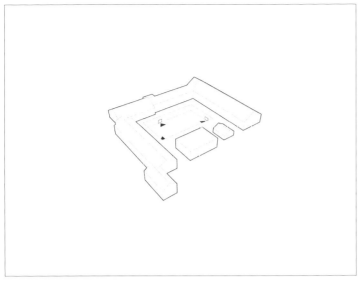

Design Concept 设计概念图

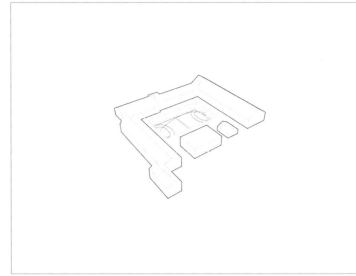

Design Concept 设计概念图

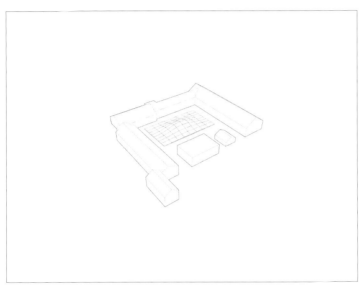

Design Concept 设计概念图

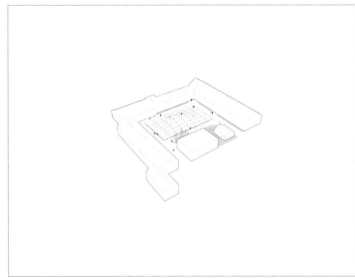

Design Concept 设计概念图

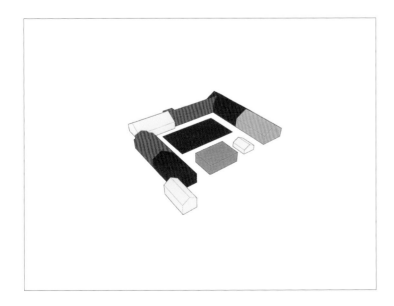

Design Concept 设计概念图

◯ Design Highlights : Prevent From Destorying the Style of Architecture

The new multi-purpose hall will primarily be for the pupils' physical education and social development. The hall is placed 5 m below ground in the centre of the school's courtyard which ensures a good indoor climate, low environmental impact and high architectural quality. The characteristic soft curved roof wood construction will act externally as an informal meeting place that can host numerous activities from group work to larger gatherings. The edge of the roof is designed as a long social bench with easy access across the courtyard and is perforated with small windows to ensure the penetration of daylight.

○ **设计亮点：地下设施避免破坏建筑风格**

新的多功能大厅主要用于学生的体育教育和发展社交活动。大厅位于地下 5 m 处，为保证大厅有良好的室内环境，减少对环境的影响，让大厅有好的建筑质量，设计师把大厅建在学校操场的中心。大厅那木制的、柔软的屋顶相当独特，学生可以在上面进行非正式的集会。

无论是一小组人,还是一大群人,都可以在这里举办各种各样的活动。屋顶的边缘被设计成方便学生社交的长凳子,能轻易地穿越操场。尾顶的边缘还被凿了不少小孔,并装上了小窗户,以方便日光照射到下面。

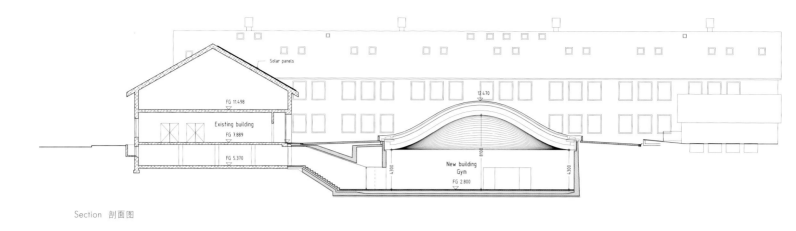

Section 剖面图

◐ **Design Details**

Solar panels placed strategically around the existing buildings provide heat for the hall. Opposed to placing the hall outside the school's building – thus spreading the social life even more – the new hall creates a social focal point and connection between the existing facilities of the high school.

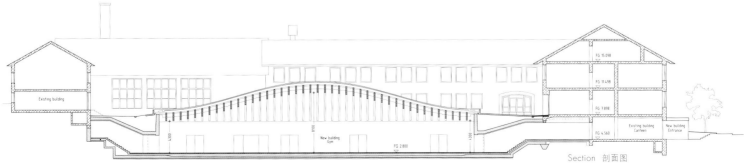

Section 剖面图

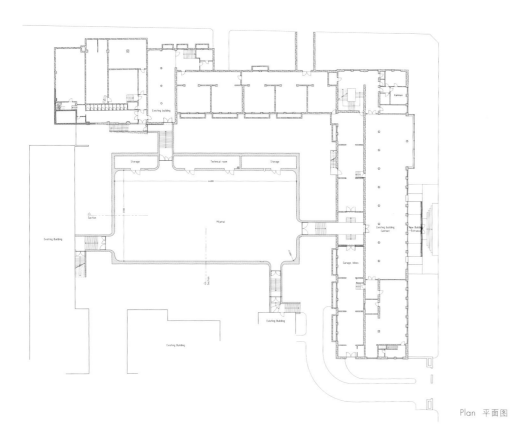

Plan 平面图

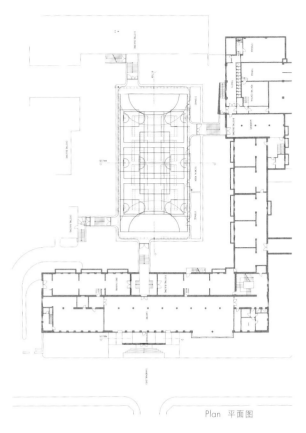

Plan 平面图

◗ 设计细节

经过战略性的考虑，设计师把太阳能电池板放置在现有建筑的周围，为地下的大厅提供热量。与把大厅设立在学校建筑的外面相比，设计师更多地从扩展社交活动方面去考虑，而新的大厅正好体现了以拓展社交活动为重点这一主题，为高级中学现有的设施创建了很好的连接。

- **Architect:** John McAslan + Partners
- **Construction:** Tata Steel Project, Arup
- **Location:** London, UK
- **Area:** 80,000 m²

- 建筑设计公司：John McAslan + Partners
- 施工单位：Tata Steel Project, Arup
- 地点：英国伦敦
- 面积：80 000 m²

国王十字火车站改建工程

Transforming King's Cross

Project Overview

As an important transportation hub in London, King's Cross Station has a long history, opening to the public in 1852. The redevelopment now accommodates 50 million passengers passing through the station every year. The project began as a concept masterplan in 1998 and construction began in 2005, once London won the bid to host the 2012 London Olympics.

项目概况

作为英国伦敦重要的交通枢纽,国王十字火车站有着悠久的历史。它于1852年建成并投入使用。改建后,每年能容纳5 000万人次的客流量。项目在1998年开始设计总平面图,在2005年开始施工,那个时候伦敦正好成功获得2012年伦敦奥运会的举办权。

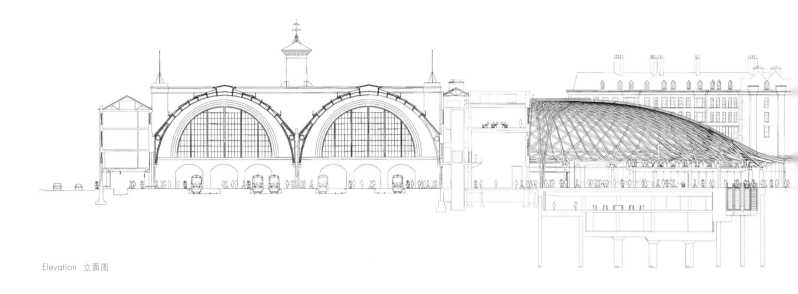

Elevation 立面图

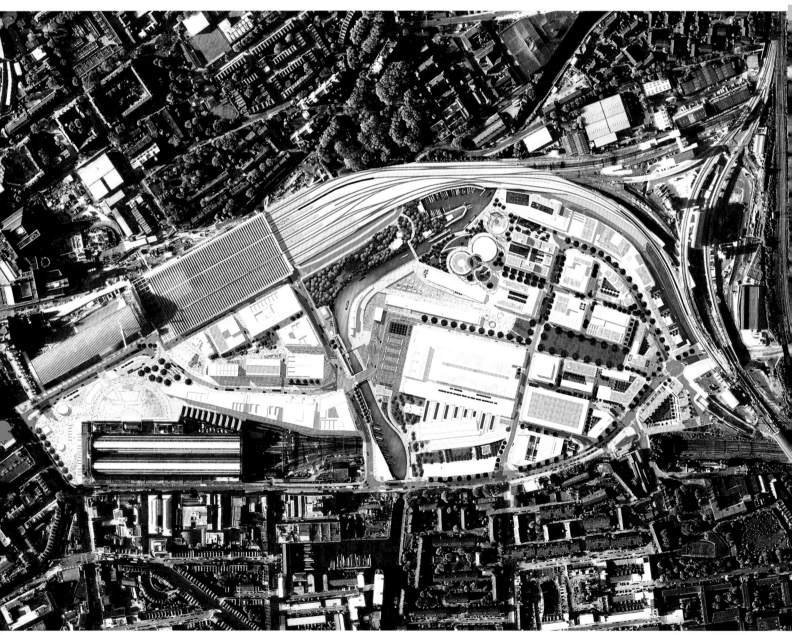

Site Plan 总平面图

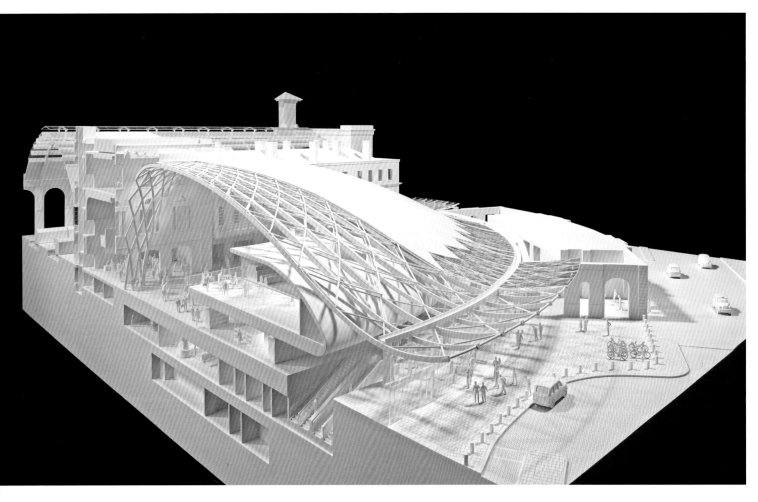

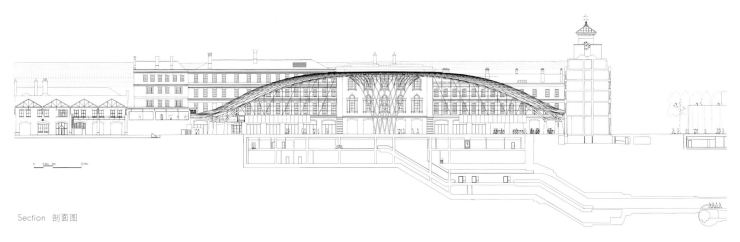

Section 剖面图

◯ Design Highlights : Re-use, Restore and Create

The transformation of King's Cross Station for Network Rail involves three very different styles of architecture: re-use restoration and new build. The train shed and range buildings have been adapted and re-used, the station's previously obscured Grade I listed façade is being precisely restored, and a new, highly expressive Western Concourse has been designed as a centerpiece and the 'beating heart' of the project. Opened to the public on 19 March 2012, King's Cross is now recognized as a new, iconic architectural gateway to the capital.

○ 设计亮点：保留、修复与新建

英国铁路路网公司对这个项目的改造涵盖保留、修复和新建等三类工程。保留了列车棚和周边建筑，并对其作调整，修复车站之前被遮挡的一级保护建筑立面，新建的西大厅极富表现力，如同"跳动的心脏"成为项目的核心。在2012年3月19日向公众开放的国王十字火车站现在已经是一个公认的首都新标志性建筑。

Our design re-orientates the station to the west, creating significant operational improvements and will reveal the main south façade of Lewis Cubitt's original 1852 station. Although the Western Concourse is probably the most visually striking change to the station, JMP's work on the project also involves a series of layered interventions and restorations including the restoration of the Eastern Range building and the revitalization of the Main Train Shed, Suburban Train Shed and Western Range buildings.

我们为车站重新定向，朝西开放使其运营得到明显改善，同时也令由路易斯·丘比特设计的在1852年投入使用的车站南侧主立面显露出来。虽然西大厅或许是整个项目最大的亮点，但我们还是进行了一系列分层介入和修复工作，包括东楼的修复，以及主线列车棚、郊线列车棚及西楼的更新。

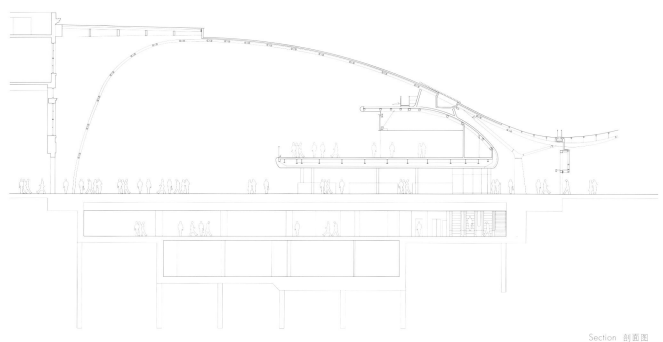

Section 剖面图

Design Details

The centerpiece of the £550 million redevelopment is the new vaulted, semi-circular concourse to the west of the existing station. The concourse rises some 20 m and spans the full 150 m-length of the existing Grade I Listed Western Range, creating a new entrance to the station through the south end of the structure and at mezzanine level to the northern end of the Western Concourse. The 7,500 m² concourse has become Europe's largest single-span station structure, comprising of 16 steel tree form columns support the loads to the perimeter of the structure, radiating from an expressive, tapered central funnel. The graceful circularity of the concourse echoes the form of the neighboring Great Northern Hotel, with the ground floor of the hotel providing access to the concourse.

设计细节

在这个总投资 5.5 亿英镑的改造项目中，最核心的部分是现有车站西侧新建的半圆拱形大厅。大厅能显现高约 20 m，横跨 150 m 的一级保护建筑西楼，大厅南端的地面层和北端夹层开设了车站的新入口。7 500 m² 的大厅由极具表现力的锥形中央漏斗状屋面辐射至周边的 16 根承重树状钢柱，成为欧洲单体跨度最大的车站建筑。优雅的半圆形大厅与相邻的大北方酒店交相呼应，从酒店首层就可以直接进入车站大厅。

Design Concept 设计概念图

The Western Concourse sits adjacent to the façade of the Western Range, clearly revealing the restored brickwork and masonry of the original station. Located above the new London Underground northern ticketing hall, and with retail elements at mezzanine level, the concourse transforms passenger facilities, whilst also enhancing links to the London Underground, and bus, taxi and train connections at St Pancras. The concourse is now the architectural gateway to the King's Cross Central mixed-use developments, a key approach to the eastern entrance of St Pancras International. It also acts as an extension to King's Cross Square, the new plaza that has been formed between the station's southern façade and Euston Road.

西大厅与西楼外立面毗连，使原有建筑修复后的砖石立面清晰呈现出来。在位于伦敦地铁北售票厅之上，并在夹层设有零售店的大厅里，设计师不仅对乘客的设施进行改造，同时增强了火车站与地铁、公共汽车、出租车和圣潘克拉斯列车线的联系。西大厅已经成为国王十字中心区多功能开发的序幕，以及圣潘克拉斯国际车站东侧入口的重要通道，也成为车站南立面和尤斯顿路之间所形成的国王十字广场的延续。

The Western Range at King's Cross is the historic station's biggest component, accommodating a wide range of uses. Complex in plan, and articulated as five buildings, the practice's considered architectural intervention has delivered greatly improved working conditions for the station staff, train-operating companies and Network Rail management teams. The Northern Wing, destroyed by bombing in World War II, has been rebuilt to its original design. The reinstatement of the Western Range also delivers key gated connections, including a new gate-line at the southern end, now the main point of connection between the Western Concourse and the platforms of the Main Train Shed.

作为这座历史悠久的国王十字火车站的最大组成部分，西楼设计了广泛的使用功能。通过对这五座相连建筑复杂的规划，从设计师深思熟虑的设计实践中，可以看到建筑的介入极大地改善了车站工作人员、列车运营公司以及英国铁路路网公司管理团队的工作条件。在第二次世界大战中被炸毁的北翼已经按原貌重建。此外，西楼的修复还为车站提供了包括位于南端的新通道等重要连接，成为西大厅和主列车棚站台之间的主要连接点。

A large quantity of the passengers imposes a huge burden for the King's Cross Station. Therefore, it is important to lead the passengers more appropriately. The Western Concourse looks like a conical funnel. From this dramatic interior space, passengers access the platforms either through the ground level gate-lines in the Ticket Hall via the Western Range building, or by using the mezzanine level gate-line, which leads onto the new cross–platform footbridge. The quantity of the passengers is both large on the ground and below ground. There is a metro station under the railway station. As one of the most important transportation, the metro trains provide a lot of passengers to the railway stations. So the underground facilities were optimized according to the flow of the passengers, ventilation, lighting, the rationality of the signs, the safety of the escalator, etc. After the optimization, the links to the underground station is more unhindered than before. The optimization plays an important role in leading the flow of the passengers. The King's Cross Station with unhindered connections, will be more and more important in the transportation.

大量的客流给国王十字火车站造成很大的负担，因此，如何更好地引导客流，成为改建工程中的一个重点。西大厅像一个锥形漏斗。在这个极富吸引力的空间内，乘客既可以从地面层入口进入售票大厅，经由西楼到达站台，也可以从夹层入口进入，经由新建的跨越站台的人行天桥到达各站台。不仅地上客流密集，地下的客流也是非常庞大。火车站的下面有一个地铁站。作为重要的交通工具之一，地铁为火车站输送了大量的人流。鉴于这一点，设计师从人流动线、通风、照明、标志牌、电梯的安全性等角度出发，对地下的设施的各个方面进行了优化，加强火车站与地铁站的联系，从而起到很好的疏导客流的作用。客流畅通的火车站将发挥着越来越重要的作用。

The station's Main Train Shed is 250 m long, 22 m high and 65 m wide, spanning eight platforms. The restoration includes revealing the bold architecture of the original south façade, re-glazing the north and south gables and refurbishing platforms The two barrel-vaulted roofs are currently being refurbished and are lined with energy-saving photo-voltaic arrays along the linear roof lanterns, while a new glazed footbridge extends from the mezzanine level of the Western Concourse, spanning the Main Train Shed and providing access to platforms 0-8 via lifts and escalators. JMP's design integrates the main and suburban train sheds for the first time, creating a coherent ground-plan for passenger movements into and through the station. Improvements to the Suburban Train Shed located to the north of the Western Concourse and Western Range buildings have enhanced the operation of its three platforms (the busiest in the station during peak-hours).

车站的主列车棚长 250 m，高 22 m，宽 65 m，横跨 8 个站台。修复工程包括展现原有建筑的南立面、重新粉刷南侧和北侧建筑的尖顶以及翻新站台。两个筒形穹顶层面正在翻新，并沿着线性的屋顶灯安装了节能光伏阵列，而新的玻璃人行桥从西大厅的夹层横跨主列车棚，并通过电梯和扶梯分别通往 8 个站台。设计师在设计中首次将主列车棚和市郊列车棚进行整合，为进入和通过车站的客流制定了流畅的路线规划。位于西大厅北侧的市郊列车棚以及西侧环形楼群的改进加强了高峰期最繁忙的 3 个站台的运行能力。

The ambitious transformation of the station creates a remarkable dialogue between Cubitt's original station and 21st-century architecture - a quantum shift in strategic infrastructure design in the UK. This relationship between old and new creates a modern transport super-hub at King's Cross, whilst revitalizing and unveiling one of the great railway monuments of Britain.

国王十字火车站改造的目标是在丘比特原有车站以及21世纪的建筑之间建立非同寻常的对话,是英国战略性基础设施设计的巨大飞跃。新旧建筑之间的关系使国王十字火车站成为现代化的超级枢纽,使这座英国最伟大的古老火车站重现生机。

- **Architects:** BNKR Arquitectura
- **Project Leader:** Arief Budiman
- **Project Team:** Arief Budiman, Diego Eumir, Guillermo Bastian & Adrian Aguilar
- **Partners:** Esteban Suárez (Founding Partner) y Sebastián Suárez
- **Photographer:** Sebastian Suárez
- **Location:** Mexico City, Mexico
- **Area:** 775,000 m²

- 设计公司：BNKR Arquitectura
- 项目负责人：Arief Budiman
- 设计团队：Arief Budiman, Diego Eumir, Guillermo Bastian & Adrian Aguilar
- 合作者：Esteban Suárez (Founding Partner) y Sebastián Suárez
- 摄影师：Sebastian Suárez
- 地点：墨西哥墨西哥城
- 面积：775 000 m²

摩地大楼

The Earthscraper

○ Project Overview

Today, the Historic Center is in desperate need of a programmatic make-over. New infrastructure, office, retail and living space is required but no empty plots are available. Federal and local laws prohibit demolishing historic buildings and even if this was so, height regulations limit new structures to eight stories. So the designers have a massive program of hundreds of thousands of square meters and nowhere to put it. This means the only way to go is down.

○ 项目概况

今天，墨西哥历史中心非常需要有计划性的转变。人们需要兴建新的设施、办公场所、零售商店和生活空间，但已经没有空地可以使用。联邦政府和地方法规禁止拆毁历史建筑，即使是这样，对建筑的高度是有限制的，新的建筑最高不超过八层。所以设计师有一个几十万平方米的大项目无处安放。这表明我们只有一种解决办法，就是把这栋楼建在地下。

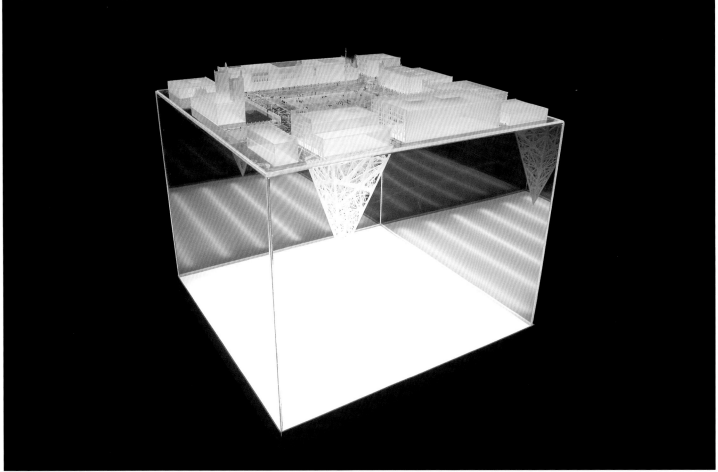

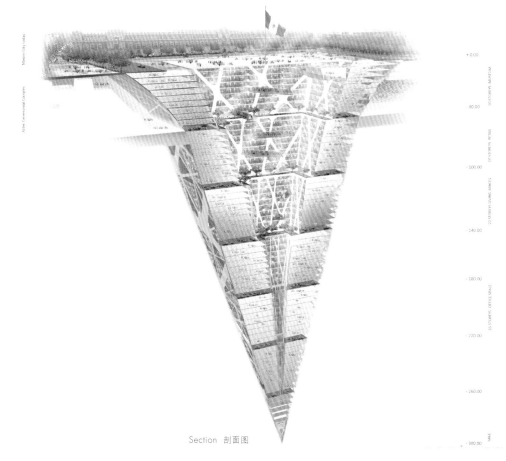

Section 剖面图

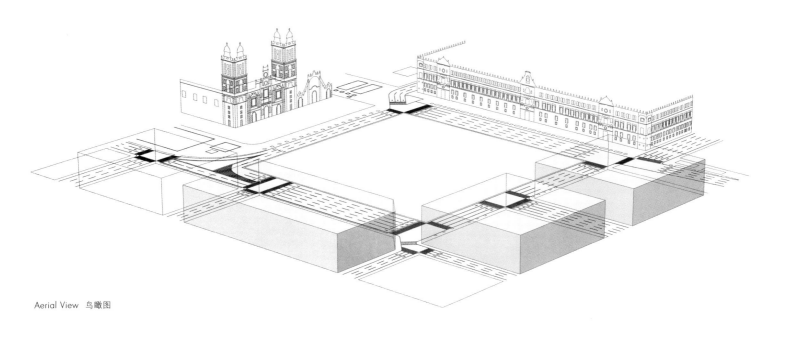

Aerial View 鸟瞰图

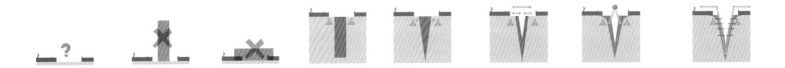

Design Concept 设计概念图

◯ Design Highlights : Preserve the Culture

The main square of Mexico City, known as the "Zocalo", is 57,600 m² (240 m x 240 m), making it one of the largest in the world. It is bordered by the Cathedral, the National Palace and the City Government buildings. A flagpole stands at its center with an enormous Mexican Flag ceremoniously raised and lowered each day. This proved as the ideal site for the Earthscraper: an inverted skyscraper that digs down through the layers of cities to uncover people's roots.

◯ 设计亮点：人文保护的空间

墨西哥的主广场，被人们称为"宪法广场"，其面积是 57 600 m²（240 m x 240 m），是世界上最大的广场之一。它与大教堂、墨西哥国家宫和市里面的政府大楼相接壤。旗杆矗立在广场中心，每天，巨大的墨西哥国旗都在这里隆重地升起、降落。这提供了设计摩地大楼的灵感。这是一座倒置的摩天大厦，它深入城市的深处，帮人们追根溯源。

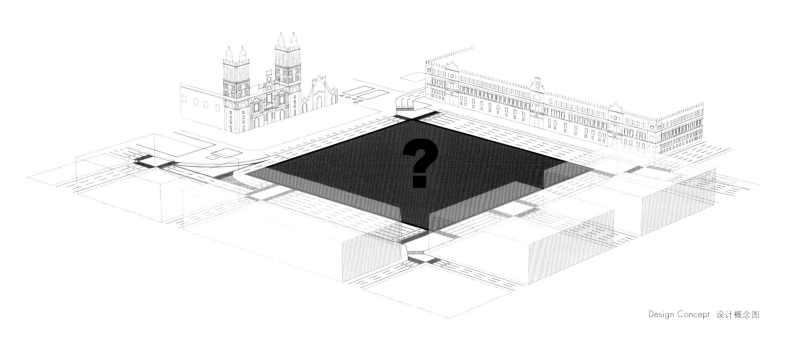

Design Concept 设计概念图

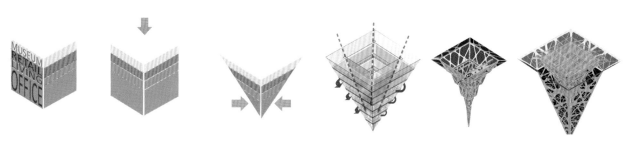

Design Concept 设计概念图

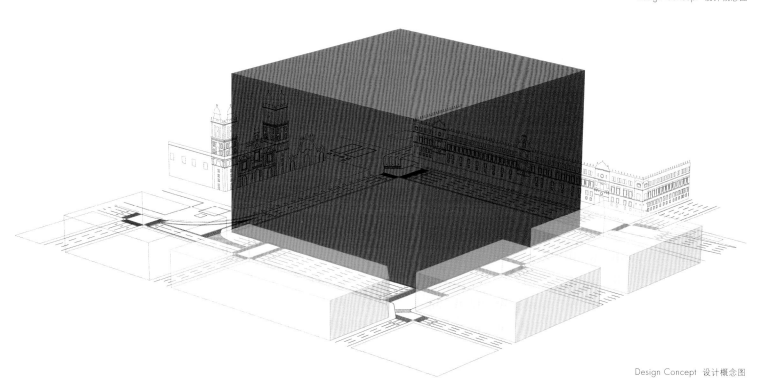

Design Concept 设计概念图

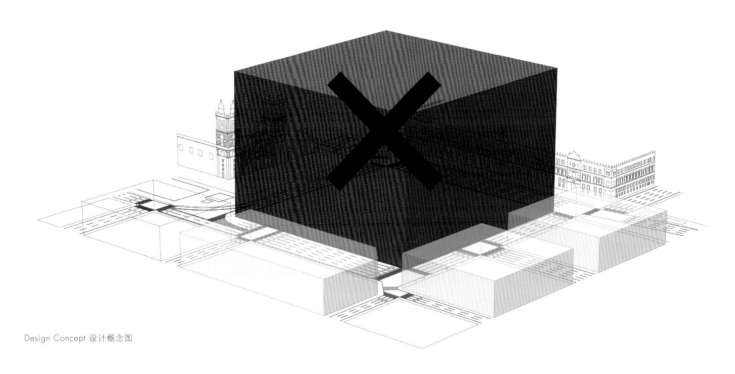

Design Concept 设计概念图

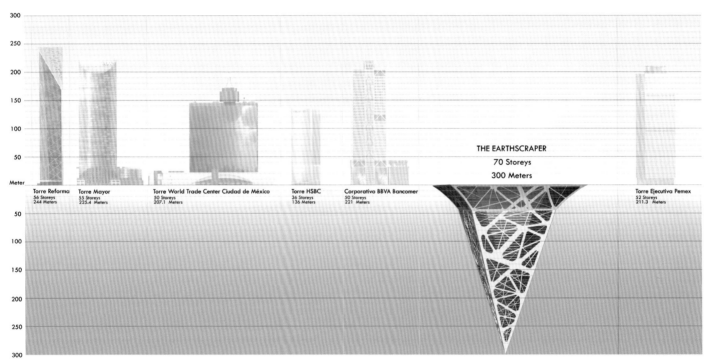

Elevation 立面图

The Earthscraper is the Skyscraper's antagonist in a historic urban landscape where the latter is condemned and the preservation of the built environment is the paramount ambition. It preserves the iconic presence of the city square and the existing hierarchy of the buildings that surround it. It is an inverted pyramid with a central void to allow all habitable spaces to enjoy natural lighting and ventilation. To conserve the numerous activities that take place on the city square year round (concerts, political manifestations, open-air exhibitions, cultural gatherings, military parades…), the massive hole will be covered with a glass floor that allows the life of the Earthscraper to blend with everything happening on top.

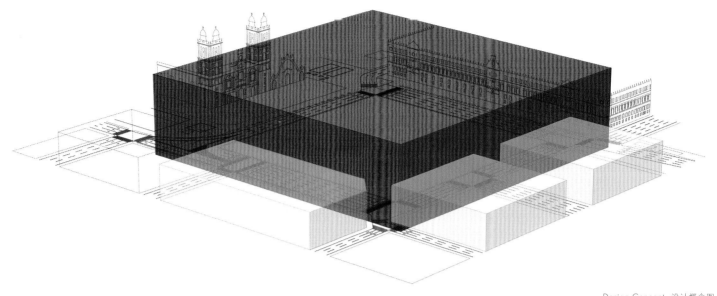

Design Concept 设计概念图

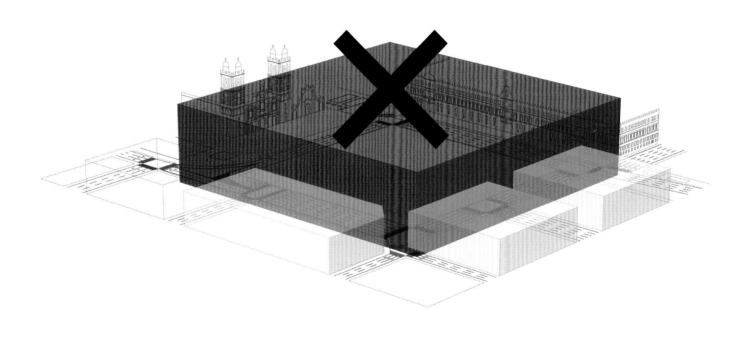

Design Concept 设计概念图

摩地大楼是一座与摩天大楼相反的建筑，在具有如此历史文化的城市景观背景中，后者被认为是不适合建造的对象，因为把它建起来会破坏古建筑。建造摩地大楼最主要的目的是保护建筑周边的环境，保留城市广场的标志和周边建筑的层次。这是一座倒置的、金字塔状的、采用中空结构的、可以让所有适宜居住的空间拥有完善的采光、通风系统的独特建筑。它可以节省一年四季所举办的大量活动所占用的地上空间（包括演唱会、政治活动、露天展览、文化交流活动、阅兵），缓解用地矛盾，地面上的开口会用一块玻璃地板盖住，以便让摩地大厦里的生活能与地上面的生活接轨。

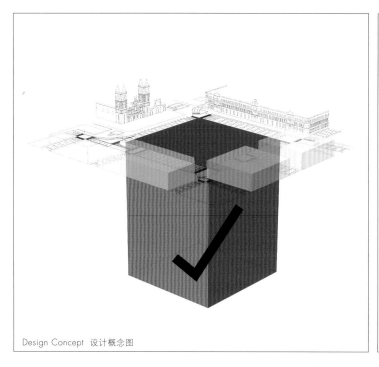

Design Concept 设计概念图

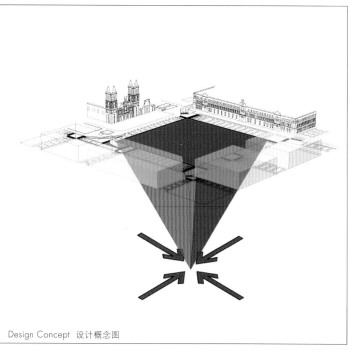

Design Concept 设计概念图

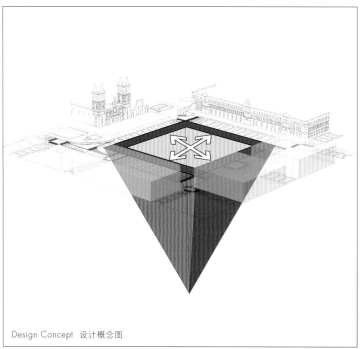

Design Concept 设计概念图

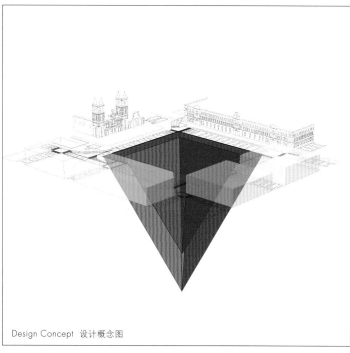

Design Concept 设计概念图

The main objective of this project is to achieve further densification of the Historical Centre. The designers consider that the benefits of densification, both environmental and cultural, are significant. Mexico City, a city of 21.2 million people, has enormous problems with urban sprawl. This sprawl causes a city which accounts for 25% of Mexico's population and 38% of its GDP to be constantly clogged with traffic which not only makes its air infamous for its pollution but also seriously decreases the quality of life of its residents. Development of the city centre is crucial to solving this problem, and the Earthscraper is simply a unique response to this need.

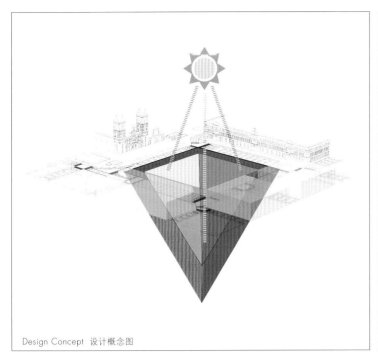
Design Concept 设计概念图

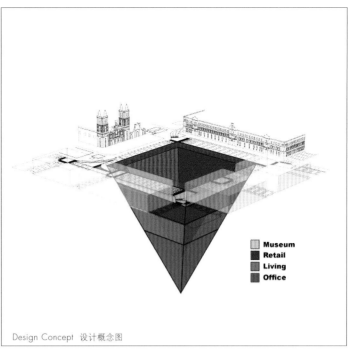
Museum
Retail
Living
Office

Design Concept 设计概念图

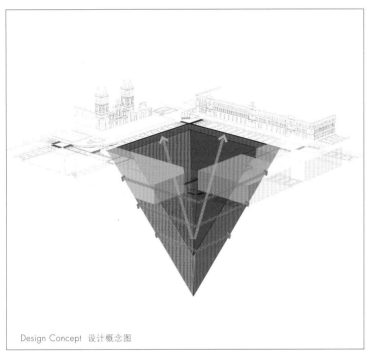
Design Concept 设计概念图

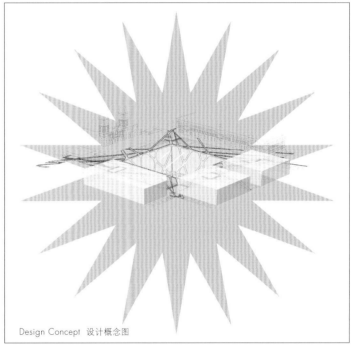
Design Concept 设计概念图

建设这个项目的主要目的是满足历史中心进一步的用地需求。设计师曾从环境和文化的角度，考虑过开发土地、满足更多的用地需求是否可行这个问题，结果发现是可行的。墨西哥城，这个有着2.12千万居住人口的城市，在城市开发上碰到不少问题。城市的开发，导致该城市25%的人口和38%的GDP不断受到交通堵塞的影响。交通堵塞不仅导致空气质量变差，而且会严重地影响到居民的生活质量。城市中心的合理发展，是解决这一问题的关键，摩地大楼只是从实际出发，切实解决这一问题的一种特别的形式而已。

○ **Design Details**

The Earthscraper is an inverted pyramid, the internal slope of which is designed to maintain the structural integrity of its wall. This greatly reduces the need for retaining structure and does not create problems for the foundations of surrounding buildings. Because its structure must already resist the lateral forces of the surrounding earth, it would be especially strong against the lateral forces of an earthquake.

○ 设计细节

摩地大厦是倒金字塔形状的，因此其内部设计了斜坡，以保证墙体结构的一致性。这大大降低了建筑维护的需要，而且不会使周边建筑的地基出现问题。这样的结构能有效地抵御周围土地带来的侧向力，因此它具有特别强的抗震能力。

There is, however, a challenge in dealing with water. The Earthscraper reaches a depth of 300 m which is 165 m below the average depth of the water table in Mexico City. This means that the bottom half of the Earthscraper must essentially be designed to hover on muddy terrain. In short, this type of building implies a greater investment in structure. One significant advantage to building underground is the thermal insulation provided by the earth. This building, while requiring excellent ventilation systems, would require less temperature control.

然而，在供水方面，有一个富有挑战性的问题需要解决。摩地大厦可以到达地下 300 m 的深度，但墨西哥城的水表在地下的平均深度只有 165 m。这意味着摩地大厦靠近底部的那一半楼层在设计上必须要保证软土坍塌的问题不会发生。简而言之，这种建筑结构需要更大的投资。把它建在地下的一个明显优势是，可以利用土地带来的隔热效应。这座建筑需要良好的通风系统，但在温度的控制方面可以少费点心。

Besides the structural challenges of such a project, it is very important that the designers have systems in place to make living underground acceptable. Among these is the solution to the need for natural light in the lowest levels. In these levels a system of fibre optic illumination would need to be put into place which would guarantee natural a light even at the greatest depths.

除了要解决结构上的这些富有挑战性的问题,设计师还有更重要的事情要做,那就是设置一些系统,使人们更喜欢地下的空间。为此,需要满足采光的最低要求,引进光纤照明系统,这可以保证自然光能照射到摩地大楼的最底下的楼层。

- **Architects:** ON-A
- **Client:** Archdiocese of Tarragona
- **Photographer:** Jordi Fernández
- **Location:** Tarragona, Spain
- **Surface:** 7,000 m²

- 设计公司：ON-A
- 客户：Archdiocese of Tarragona
- 摄影师：Jordi Fernández
- 地点：西班牙塔拉戈纳
- 建筑表面积：7 000 m²

塔拉戈纳神学院改建

Seminary Reform

○ **Project Overview**

The project focuses on the conversion of the Seminary of Tarragona into a culture space and a landmark for the city. The new prestigious research centre contains science space, art space, religion space and history space. These spaces are part of the essence of the project. The clear intention of reforming the seminary is to open the building to the public, to discover all its rooms and show the world a unique cultural heritage that is now hidden.

○ **项目概况**

项目的重点是将塔拉戈纳神学院改造成一个文化空间和城市地标。这个新的著名的研究中心包含科学空间、艺术空间、宗教空间和历史空间。这些空间都是项目的精华。改建的目的非常明确，设计师希望向公众开放参观这座建筑，向人们展示它的全貌，向世界展示这个目前处于隐蔽状态的、独特的文化遗产。

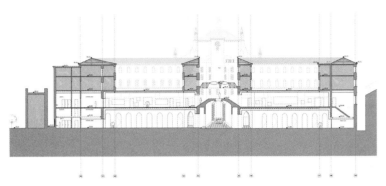

Section 剖面图

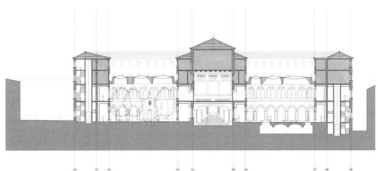

Section 剖面图

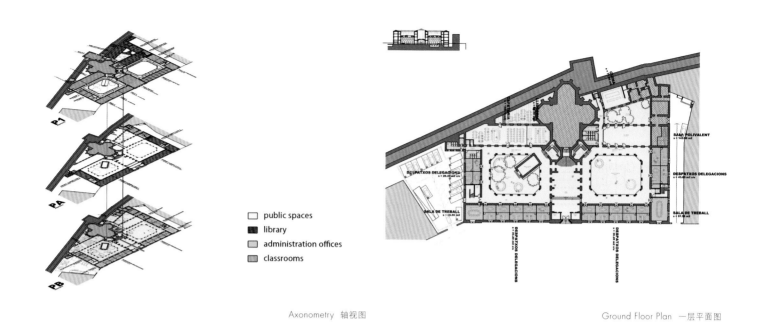

☐ public spaces
■ library
☐ administration offices
☐ classrooms

Axonometry 轴视图

Ground Floor Plan 一层平面图

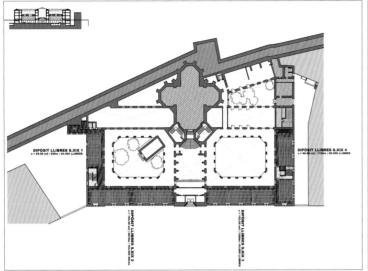

Mezzanine Plan 夹层平面图

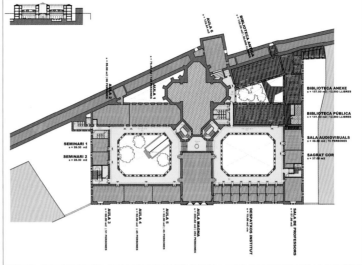

First Floor Plan 二层平面图

○ Design Highlights : Modern Technologies Hosting all Sorts of Culture

The designers provide the building with the most modern technologies to host all sorts of cultural events but with great respect of the historical environment, preserving it and emphasizing it. There is a religion space with strong characteristics. Some circular glasses of different sizes were set in the roof. The nature light shines into the hall through the circular glasses. That makes the hall bright enough. The ceiling was divided into some lattices of different sizes by some lines. The interior space is surrounded by the walls which contains some arched doors. The doors on the first floor are not as big as the doors on the ground floor. There is also a church that contains some arched doors in the middle of the hall. The whole space is very harmonious. A smaller room is under the hall. The underground room is a unique room with a special roof which contains four arched surfaces. Three beams attached the roof and the floor, which makes the room become stable. The room is filled with ancient Rome characteristics, which perfectly match the hall. Traveler can fully enjoy the history and culture of Rome here.

○ 设计亮点：用现代技术展现建筑文化

设计师利用最现代的技术，让建筑的各种文化元素得以最充分的展现，而且充分尊重建筑的历史环境，保留并突出这些能体现历史文化背景的设施。这里比较有特色的是一个宗教空间。屋顶镶嵌着一些大小不一的圆形玻璃，日光透过圆形玻璃，照进室内，保证室内有足够的光线。天花板被一些线条分成大小不一的格子。室内四周是一些有着不少拱形门的墙壁，墙壁有两层，上层的门比较小，下层的门比较大。中间有一座古罗马风格的教堂。整个空间搭配得非常和谐。这个大厅下面有一个面积较小的房间。这个地下的房间造型非常独特。房顶由四个拱形的面组成，有三条梁柱支撑着屋顶和地面，显得非常坚固。整个房间富有古罗马风情，与地上大厅的风格相匹配。游客们将在这里充分领略到罗马的历史文化。

- **Architects:** *Oscar Tusquets Blanca*
- **Project Architects:** *Pep Palaín, Esther Villanueva*
- **Construction Engineer:** *Robert Ayala , Gerardo Barrena, Maria Roger , Susana Pavón*
- **Photographer:** *Rafael Vargas, Oscar Tusquets*
- **Location:** *Barcelona*

- 设计公司：*Oscar Tusquets Blanca*
- 项目设计师：*Pep Palaín、Esther Villanueva*
- 建造工程师：*Robert Ayala、Gerardo Barrena、Maria Roger、Susana Pavón*
- 摄影师：*Rafael Vargas、Oscar Tusquets*
- 地点：巴塞罗那

Palau de la Música
音乐厅改造

Palau de la Música Reform

◯ **Project Overview**

The Palau de la Música was built by Domènech i Montanor in 1906. So far, it is one of the most delicate and elegant concert halls in the world. It is a fantastic concert hall that combines with several elements including Spanish element and Arab element with rich-decorated façade. Oscar Tusquets Blanca had transformed the concert hall for two times in order to stabilize the building and enlarge the function of the building.

◯ 项目概况

Palau de la Música 音乐厅由 Domènech i Montaner 于 1906 年建造完成，是目前世界上最精致典雅的音乐厅之一，其立面装饰丰富，而且结合了多种元素，包括传统的西班牙元素和阿拉伯元素。Oscar Tusquets Blanca 曾先后两次对其进行改造，以便让建筑更稳固，功能更完善。

○ Design Highlights: Restore and Keep the Characteristic of the Old Building

In the first time, the designers restored the very damaged parts of the old building and equipped it with sanitary facilities, accessibility, safety and comfort. The most significant part that they do was take advantage of part of the site of the adjacent unfinished church to open up the central well. In the extension adjacent to the stage they provided services for the performers: rehearsal room, dressing rooms, library, etc.

Elevation 立面图

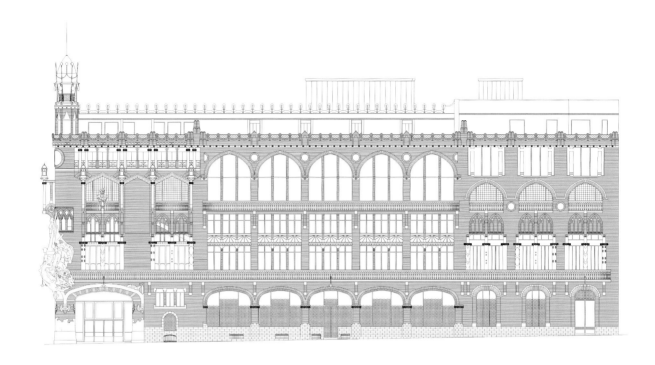
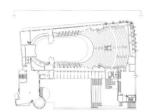

Elevation 立面图

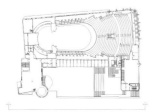

Elevation 立面图

◯ **设计亮点：修复旧建筑，保留原有特色**

在第一次改造中，设计师修复了旧建筑受损严重的部分，设置了配套的卫生设施，提升了建筑的可达性、安全性和舒适度。他们所做的最重要的工作是利用旁边未完工的教堂的地理位置的优势，开辟中央天井。为了给演员提供方便，在连接舞台的地方，设计师扩建了排练室、更衣室和图书馆等设施。

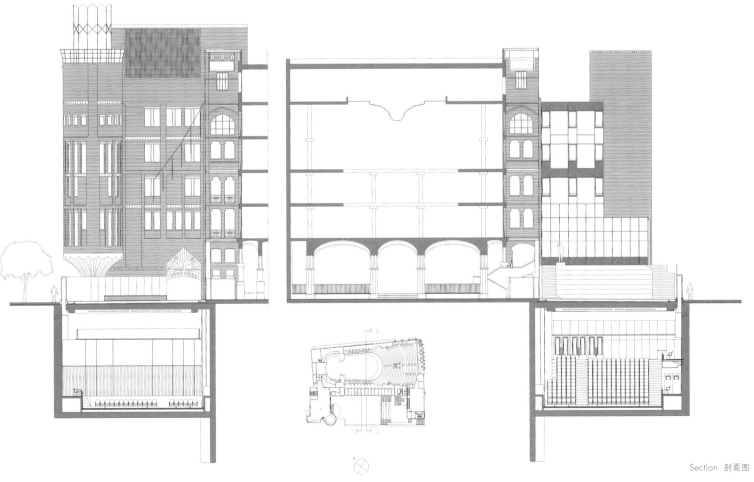

Section 剖面图

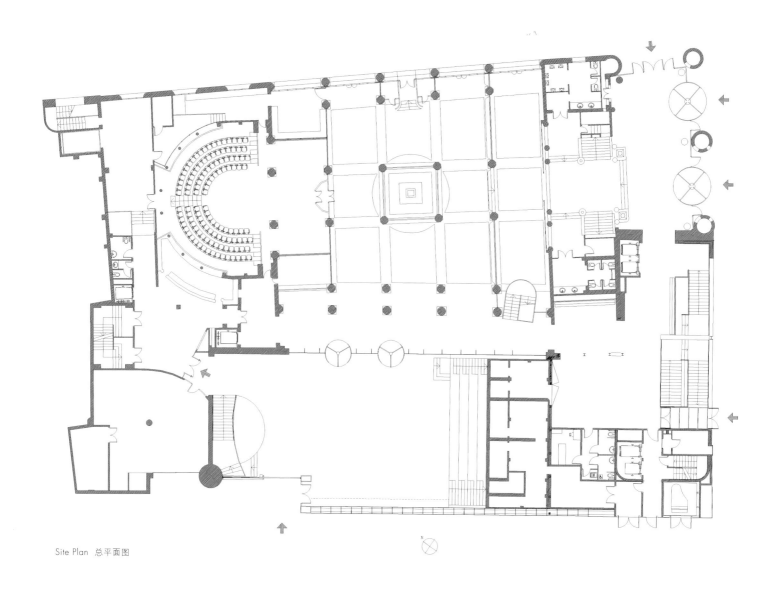

Site Plan 总平面图

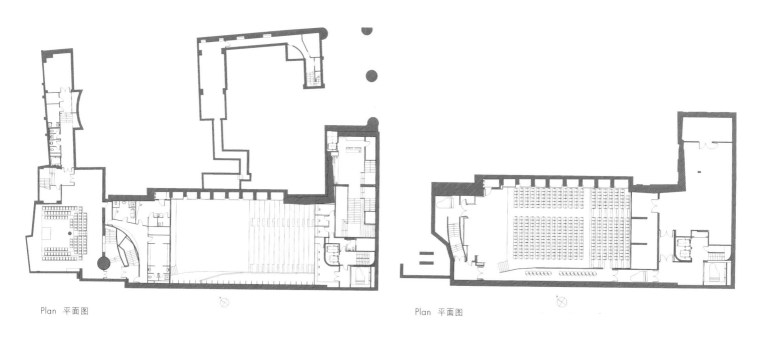

Plan 平面图

Plan 平面图

○ **Design Detail**

The second transforming is related to the newly-created underground plaza. The underground plaza an auditorium for 540 people equipped with the advanced optics system and acoustics system. It can provide the audience an unusual experience. The designers give the Palau a new façade by completely opening up richly-decorated side patio to the newly-created underground plaza. On the north side of the plaza they constructed a building for the musicians (dressing rooms, library, etc.), on the south they created one for the public and the music school. It is more convenient for people to enter or exit the plaza.

○ 设计细节

设计第二次改造工作与新建的地下大厅有关。这个大厅能容纳540人，具有先进的灯光系统和播音系统，能带给观众不同寻常的体验。设计师从天井有华丽装饰的那一边开凿，为新建的地下大厅创建一个入口，并在此基础上为 Palau 音乐厅创建的外墙。设计师在大厅的北侧为音乐家建造一座建筑（包含更衣室、图书馆等），在大厅的南面为公众及音乐学校建造一座建筑，这样更方便人们进出音乐厅。

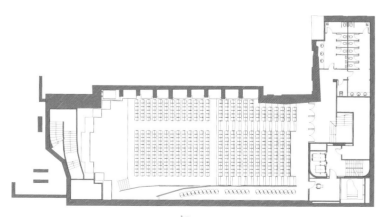

Plan 平面图

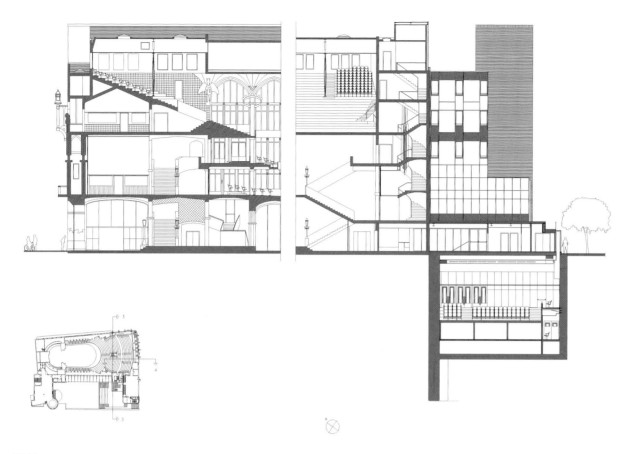

Section 剖面图

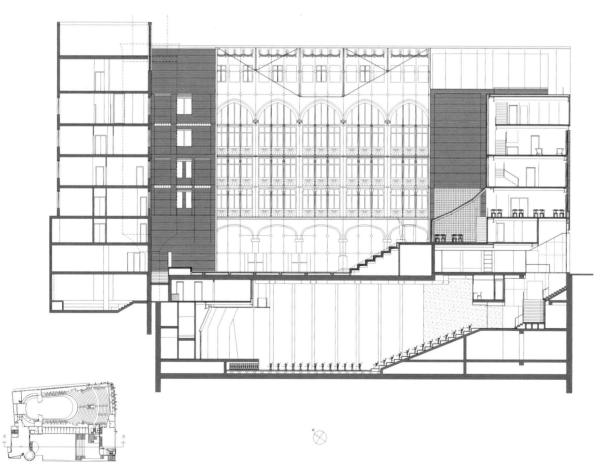

Section 剖面图

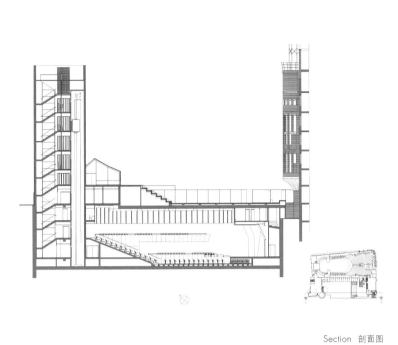

Section 剖面图

- **Architects:** *Karres en Brands LR*
- **Design Team:** *KBP.EU, a joint venture between Karres en Brands and Polyform*
- **In Collaboration with:** *Oluf Jøgensen Engineering and Ulrike Brandi Licht*
- **Client:** *Municipality of Copenhagen*
- **Location:** *Copenhagen, Denmark*
- **Area:** *22,000m² (Center for Renhold 1,000 m²)*

- 设计公司：*Karres en Brands LR*
- 设计团队：*KBP.EU, a joint venture between Karres en Brands and Polyform*
- 合作者：*Oluf Jøgensen Engineering and Ulrike Brandi Licht*
- 客户：*Municipality of Copenhagen*
- 地点：丹麦哥本哈根
- 面积：22 000m²（Renhold 中心：1 000m²）

Kømagergade
购物步行街

Købmagergade

◯ **Project Overview**

The curved course of the Købmagergade shopping street is typical of the city centre of Copenhagen. Together with the Hauser Plads, Kultorvet and Trinitatis Kirkeplads, this long street embodies the characteristic image of the labyrinthine medieval city centre.

◯ 项目概况

Købmagergade 购物步行街位于丹麦哥本哈根市中心，是典型的哥本哈根式街道。该购物街连同 Hauser 广场、Kultorvet 广场和 Trinitatis 教堂广场，呈现出犹如中世纪般风情的城市迷宫。

Site Plan 区位图

Design Highlights: Material with Strong and Harmonious Appearance

The district has its own daily and weekly rhythms: people cycle, walk, shop, play and go out in the evenings. But traffic for deliveries, refuse collection and maintenance also joins in this rhythm. The first step of the project was to clean and empty the area, so that the flow of people can easily find its way. The designers also selected strong materials such as natural stone: a durable material with a strong and harmonious appearance. The design proposal encourages the development of intensive city life on the one hand, and on the other it is linked with the rich history of Copenhagen.

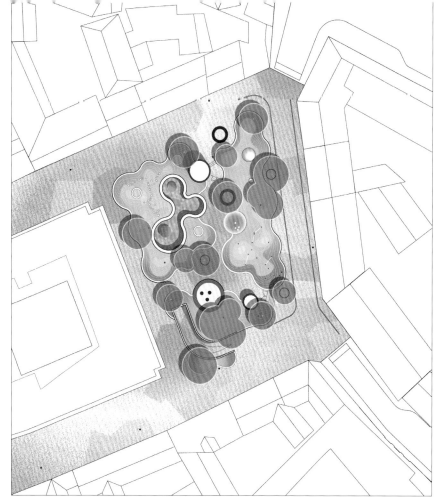

Site Plan 总平面图

设计亮点：材质与原建筑和谐搭配

地区每日和每周都有着属于自己的旋律：这是由人流、步行、购物、娱乐以及夜晚的步行街谱出的乐章。当然，货运交通、垃圾回收和交通维护也共同奏响了这一曲。该项目的第一步是对于空间的清理和重整，这样更便于行人找到路线。设计师还选择利用坚固的材料（如天然石材）来进行改造，这样的材质能够与原本的建筑外观和谐搭配。为了适应密集的城市生活，这项改造工程一方面鼓励城市的自由发展，另一方面也尽量地维护哥本哈根悠久的历史风貌。

The layout of the three squares is varied, just as their historical situation and their location in the city are varied. On the Kultorvet the dark – almost black – paving pattern of the stone is inspired by the 18th century coal trade. On the rather more peaceful Hauser Plads square, the exciting grass play mounds form a green oasis in the urban fabric. At night, the Trinitatis Kirkeplads with its famous observatory Rundetarn is transformed by artificial lighting into an enormous starry sky. The three squares are diverse in colour, from dark coal to bright stars: 'From Kultorvet to the Milky Way'.

三个广场的装修与布局是多样的，因为它们在城市中的历史背景和地理位置不同。Kultorvet 的广场地面上的石头铺装，是受到18世纪煤炭贸易的启发的，因此，这里所展现的是"黑暗"，几乎是完全的黑暗。Hauser 广场更为平易近人，这里是一座令人兴奋的草丘，人们可以在上面尽情玩耍，同时这里也成为城市绿色系统的一部分。因此，在这座广场上，设计师们利用人工照明，模拟了广袤的星空。三座广场从完全黑暗到璀璨的星空，模拟出由黑洞到银河系的太空景象。

In the evening and at night, the medieval city has its own melancholy and mysterious atmosphere, especially in winter. This unique ambience is emphasised by the use of warm indirect lighting, with a few extra accents on the squares. This means that it is still possible to see the stars, just as Christian the Fourth did from the observatory Rundetarn in the 17th century.

在傍晚和夜间，尤其是冬季，这座从中世纪遗留下来的古城会散发出属于自己独特的忧郁和神秘的气氛。这种独特的气氛，格外需要温暖色调的间接照明方式来烘托。因此，在这里，人们仍然能够仰望星空，重现 17 世纪时，在基督教的天文台中发生的一切。

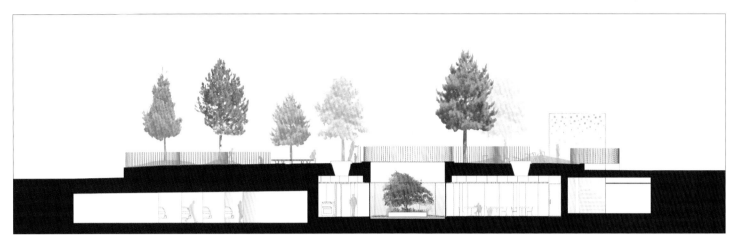

Section 剖面图

○ Design Details

Integral part of the re-design of Hauser Plads was the transformation of a former underground car park into the Center for Renhold, the headquarters of the municipal cleaning service. A patio on basement level created an outdoor space that connects with the public space on Hauser Plads.

○ 设计细节

设计师对 Hauser 广场的一部分进行重新设计，将地下的停车场改建为 Renhold 中心，也就是这个城市清洁服务的总部。设计师在地下的天井创建了一个连接 Hauser 广场公共空间的户外空间。

- **Architects:** Paolo Venturella
- **Team Design:** Paolo Venturella, Angelo Balducci, Luca Ponsi, Paolo Gaeta
- **Photographer:** wemage
- **Location:** Pristina, Serbia

- 设计公司：Paolo Venturella 建筑事务所
- 设计团队：Paolo Venturella、Angelo Balducci、Luca Ponsi、Paolo Gaeta
- 摄影师：wemage
- 地点：塞尔维亚普里什蒂纳

太阳能清真寺

◯ Project Overview

The project aim is to create a monumental and iconic building for the city of Prishtina. It is a monolithic building that becomes an urban fulcrum for the Dardania neighborhood, in the south of the city.

◯ 项目概况

建造这个项目的是为城市普里什蒂纳创建一个有纪念意义的和标志性的建筑物。这个巨大的建筑物位于城市的南部，将成为达尔达尼亚附近城市的支点。

Site Plan 总平面图

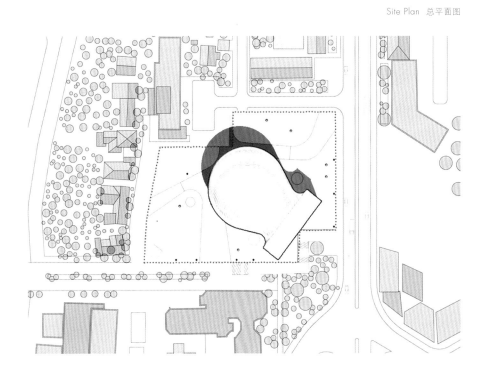

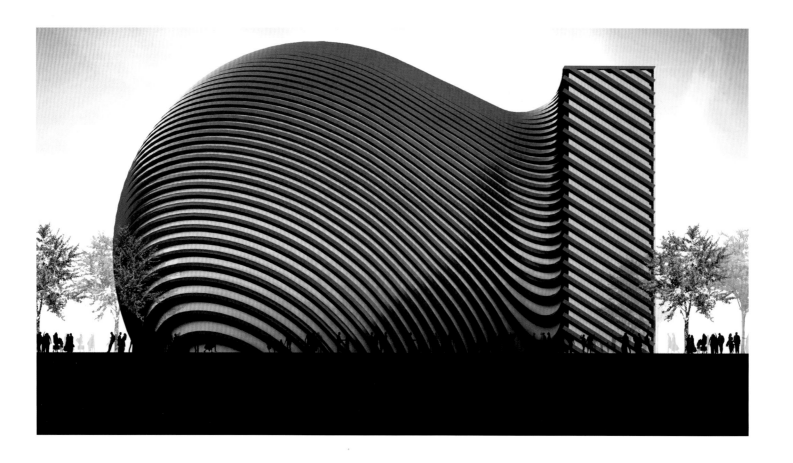

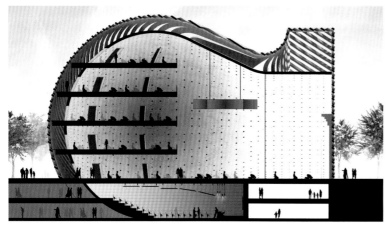

Section 剖面图

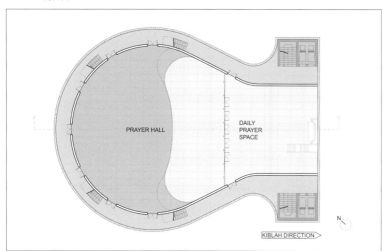

Layout 布局图

◯ Design Highlights: Elements of "Kiblah Wall" and "Dome"

The idea is to merge the two main elements of the mosque: the "Kiblah Wall" and the "Dome". The first indicates the direction where to pray and the second creates the huge space where people feel in the same community. From a geometric point of view the mosque can be seen as a sphere that comes out from the wall, but at the same time as the community facing the Mecca. This represents the link between the divine element and the prayers.

○ 设计亮点:"朝拜墙"和"圆屋顶"

项目的设计理念是融合清真寺的两个重要元素:"朝拜墙"和"圆屋顶"。前者为前来朝拜的人们指明了方向,后者为人们创造一个巨大的、统一的空间。从几何的角度来看,清真寺可以被看作是一个从墙里冲出来的球体,但同时它也是一个面向麦加的社区。这展示了代表神圣的元素与祷告之间的关系。

The prayer hall is divided into two separate rooms in order to organize a smaller area to pray from Monday to Thursday and a bigger one to fit more prayers during Friday and festivals. Both spaces are designed on multiple levels to accomodate men and women at different levels. The access is at the underground level where ancillary spaces are placed and the entrance to the Mosque is at the ground floor. In the rest of the underground areas are educational, social, administrative and commercial areas, separated from the sacred space above.

祷告大厅分成两个独立的区域,这样就可以让人们从周一到周四在较小的区域祈祷,周五和节假日在较大的区域祈祷。两个区域均分为上下多层,男士和女士可在不同的楼层进行祷告。出口位于地下层,这里有一些辅助性的空间,而清真寺的入口位于第一层。地下剩余的区域是教育、社交、行政和商业区域,与上层神圣的祷告空间隔离。

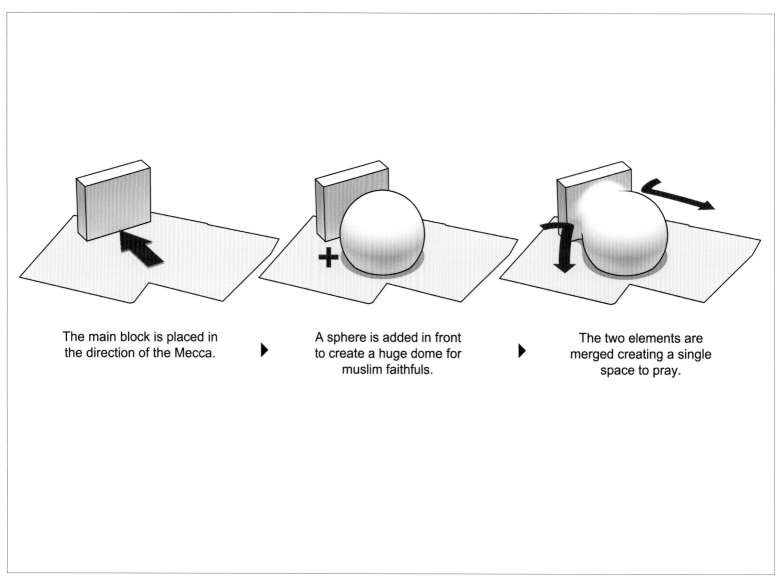

Design Concept 设计概念图

◯ Design Details

The shape of the whole envelope is designed as a "double-skin" that generates a circular space all around the prayer hall. In this space are circulations and the technological energy solutions. The exterior skin is made by a series of louvers covered with a thin photovoltaic film system that harvest energy for the use of the mosque and the other services. Taking advantage of the spherical shape, these solar panels face the sun rays during the day from the morning to the end of the day. The curved form guarantees a surface perfectly oriented to the sun during all the day so to optimize the radiation of the sun. These tilted shaders avoid the direct passage of sunlight creating indirect illumination and allow the view to outside. The project also uses a passive energetic approach thanks to its orientation. The Kiblah wall, that is south oriented, works as a huge greenhouse element that captures heat and releases it to the interior when needed in cold periods. Since the project requires four levels of underground parking, the designers also designed a geothermic energy system that provides the underground parking for either heating or sanitary purposes.

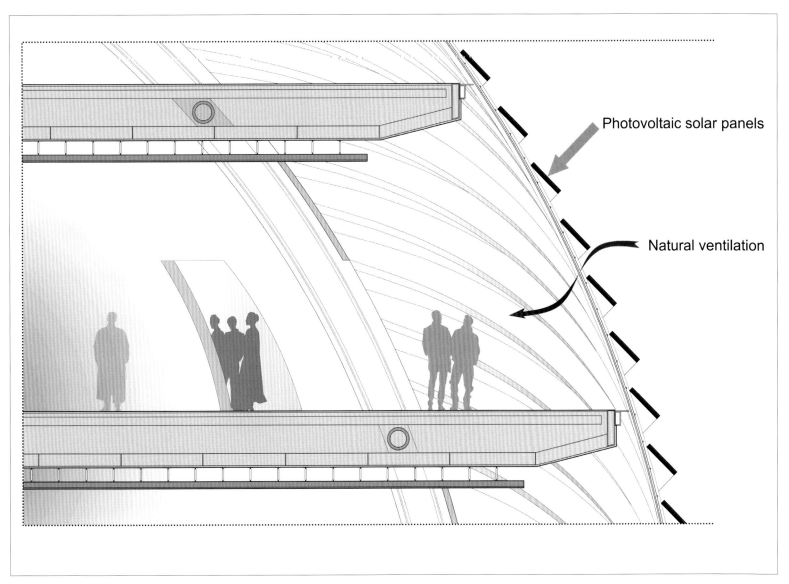

Design Concept 设计概念图

○ 设计细节

设计师将清真寺的整个外表设计成"双表皮"结构，以便在祷告大厅周围建立一个球形的空间。这个空间有不少球形的结构和为解决能源技术问题而设的设施。外表皮由一系列遮光栅格构成，栅格平面上覆盖着太阳能光伏薄膜，可为清真寺和其他公共设施收集能源。清真寺的球形设计，使得阳光可以从早到晚最大程度地照射到整个建筑。弯曲的建筑形式，能保证外表皮整天都可以面向太阳，以便以最好的方式吸收太阳辐射。成角度的遮光栅格防止阳光过度直射的同时保持了对外的开阔视野。根据其朝向太阳的方向，项目采用一个被动吸收能量的方案。朝拜墙朝向南面，可作为热质量体，在天气较冷时吸收外界热量并释放到室内。项目需要四层地下停车场，设计师为这些停车场设置了有加热或清洁消毒功能的地热系统。

- **Architects:** *Collingridge and Smith Architects (CASA)*
- **Project Architects:** *Phil Smith, CASA*
- **Design Team:** *Phil Smith, Graham Collingridge, Grayson Wanda, Chloe Pratt*
- **Structural Engineers:** *McNaughton Consulting Engineers*
- **Client:** *Ngati Hine Health Trust*
- **Location:** *Kawakawa, Northland, New Zealand*

- 设计公司：*Collingridge and Smith Architects(CASA)*
- 项目负责人：*Phil Smith、CASA*
- 设计团队：*Phil Smith、Graham Collingridge、Grayson Wanda、Chloe Pratt*
- 结构工程师：*McNaughton Consulting Engineers*
- 客户：*Ngati Hine Health Trust*
- 地点：新西兰北岛卡瓦卡瓦

Te Mirumiru
儿童早教中心

Te Mirumiru
Early Childhood Centre

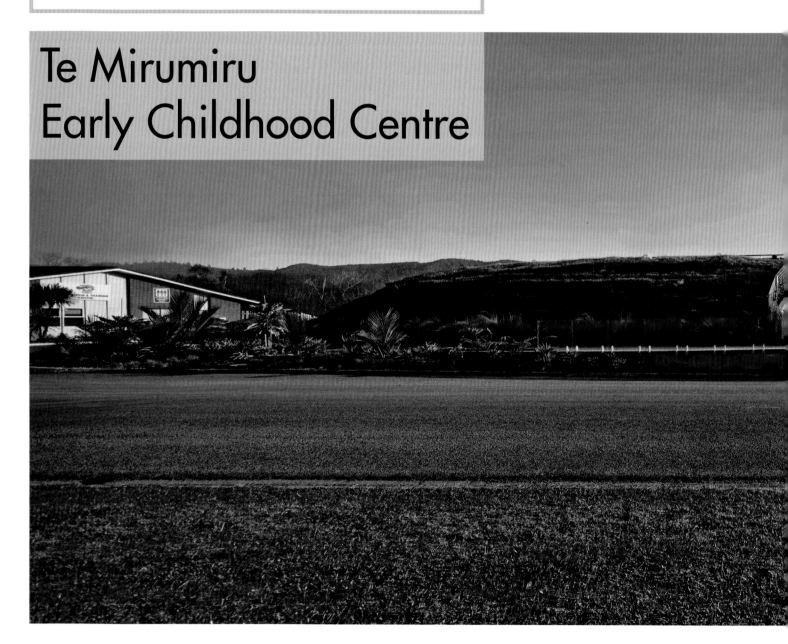

○ Project Overview

This design is an early childhood building for a Maori tribe in Kawakawa, New Zealand. The building is a place which would not only accommodate the clients children but also let the children learn about their culture and customs on a daily basis whilst having a minimal impact on the environment.

○ 项目概况

这是一个位于新西兰卡瓦卡瓦的为毛利部落而设的儿童早教中心。这座建筑不但能适应客户——孩子们的需要,而且在这里,孩子们在日常的基础训练中学习到他们的文化和风俗,与此同时,建筑对环境造成最小的影响。

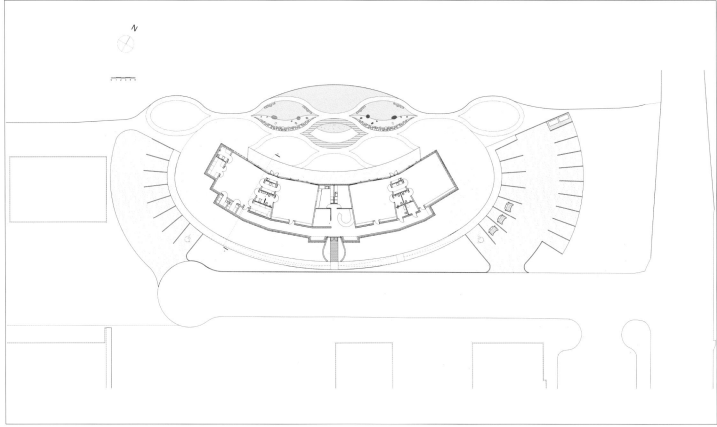

Site Plan 总平面图

Design Highlights :Tribal Cance of the Maori

The design concept for the building is based on the Maori tradition that all life is born from the womb of earth mother, under the sea. The building is located on marshy ground, with the 'womb-like form' appearing as an island, relating back to the tradition that all land is born from under the sea. A bridge is formed to give access to the island, which is symbolically shaped into the tribal canoe, representing the journey of the tribes forefathers from Hawaiki to Aotearoa (NZ).

设计亮点：毛利人部落的独木舟

毛利人认为所有生命是从地球妈妈的子宫里出生的，是从海底出生的，这座建筑恰恰体现了这个传统的思想。它被建在湿地上，其形状像母亲的子宫，也像一个小岛，它向人们诠释了一切的陆地都是从海底里长出来的这个理念。有一座桥连接着这座岛状的建筑，它看起来就像部落的独木舟，象征着部落祖先从 Hawaiki 到 Aotearoa 的旅程。

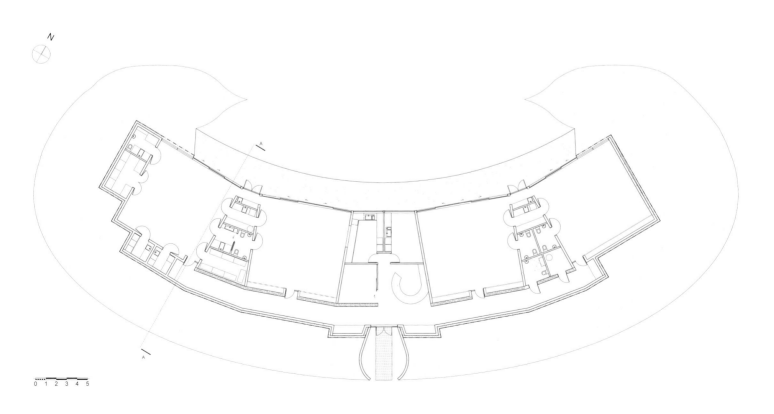

Plan 平面图

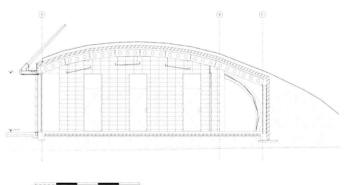

Detail 细节图

The interior, below the earth, represents the nearby Waiomio caves where the ancestors lay buried and the fortification where the ancestor Kawiti cleverly used underground shelters as defence from attack. The circular form of the design also draws inspiration from traditional fortification.

在地下的室内空间中，最具代表性的是附近埋葬毛利人祖先的Waiomio洞穴以及毛利人祖先用来抵御外来侵袭的地下避难所。设计师从传统的防御工事中得到启发，设计出这种环形的建筑设计形式。

○ **Design Details**

It was equally important to integrate passive environmental design features into the building, so all 'symbolic' features have many environmental purposes: all glazing is oriented to the north for maximum solar gain, whilst the super-insulated earth roof results in minimal heat loss, which is further assisted by the unheated circulation space placed to the south. For further internal comfort, exposed concrete construction and natural ventilation allows the building to be passively cooled in summer, with minimal heating back-up in winter provided by a solar hot water underfloor system. All spaces are naturally daylit and will need no additional electrical lighting during the daytime. All blackwater is treated on site and the clean nutrient rich water is used to irrigate the green roof. The building has been submitted for a Green Star rating and is anticipated could achieve 6 stars.

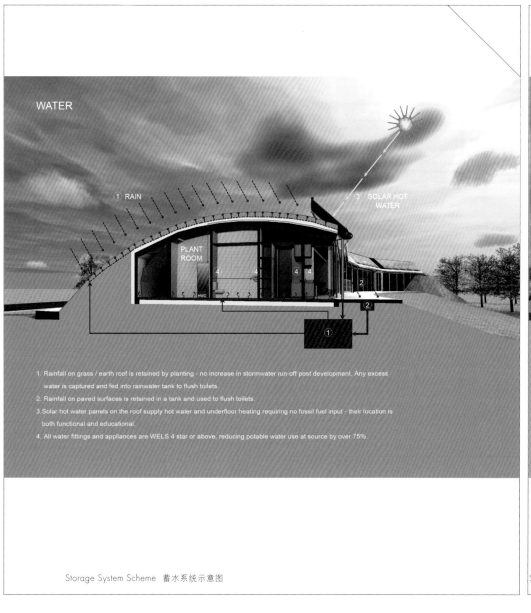

Storage System Scheme 蓄水系统示意图

Summer Cooling and Ventilation Scheme 夏天冷却和通风示

○ **设计细节**

在建筑里设置能自然调节气温，改善环境的系统同样重要。所有这些独特的建筑系统，在改善环境方面有不少作用：所有的玻璃板都面向北方，以获取最多的太阳能，同时，具有极佳隔热效果的屋顶使热量的流失降到最低值。为了使这一效果更加明显，设计师让不加热的环形空间朝向南面。为了让建筑空间更加舒适，设计师采用了暴露的混凝土建筑方案，设置了自然通风系统，这样可以让建筑在夏天自然地冷却下来，在冬天，建筑不必储存大量的热量因为地下的太阳能热水系统能提供热量。所有的空间都可以自然地采光，而且在白天，人们无需开电灯。所有的污水都是现场处理的，相对干净的、营养丰富的污水用于浇灌绿地。该建筑已经进行绿色环保评级，预计能够达到6星级。

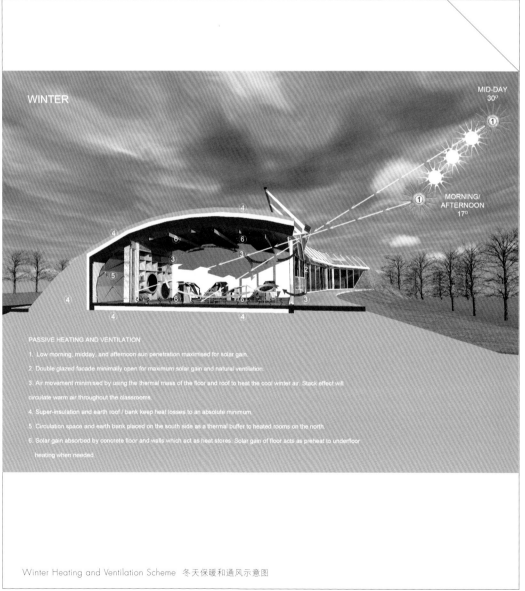

Winter Heating and Ventilation Scheme 冬天保暖和通风示意图

Chapter Three
第 3 章

The combination of underground space and urban space is completed mainly by the entrance and exit. The main function of these settings are to lead People, conduct fire evacuation, ventilate and drain contamination. The criteria of the design is judged on the rationality, safety, convenience and efficiency. A perfect underground space should meet the needs of all kinds of functions while it is attractive enough for the passengers. Hence, the value of underground space can be improved by leading a large number of people. The cases in this chapter feature the reasonable solution of traffic orientation, coordination of underground space and above ground to ensure the integrity of functions.

The Space with the Function of Leading People

引导人流的空间

地下空间与城市空间的结合主要是通过出入口空间的设置来完成的。而它的主要功能，是引导人流进出、消防疏散、通风排污等。出入口空间能否合理、安全、便捷、高效地组织人流进出是决定了地下空间出入口设计成败的关键。一个完善的地下空间出入口应该在满足各项使用功能要求的同时对行人产生足够的吸引力。因此，地下空间可以通过出入口引入大量人流以提升其价值。本章所选择的案例，既合理简明的解决了交通导向，又在时空维度上协调地下空间与广场，保证两者功能的完整性。

- **Architects:** Zechner & Zechner ZTGmbH
- **Projekt Leader:** DI Kai Uwe Preissl
- **Client:** Holding Graz Linien
- **Photographer:** Thilo Härdtlein, Helmut Pierer
- **Location:** Graz, Austria

- 设计公司：Zechner & Zechner ZTGmbH
- 项目负责人：DI Kai Uwe Preissl
- 客户：Holding Graz Linien
- 摄影师：Thilo Härdtlein、Helmut Pierer
- 地点：奥地利格拉茨

格拉茨Europaplatz广场交通枢纽

Local Transport Junction at Europaplatz Graz

○ **Project Overview**

After two years' construction, and with an Investment of 90 million euros, the local transport hub at Graz Main Station is to be finished on time and within budget. The area in front of the station has been redesigned with a new projecting roof, called 'Golden Eye' by the locals, marking the centre of the plaza. The tram lines have seen the most modernisation, and now run underground, with all four lines directly connecting with the station. Due to the Local Transport Hub offering international connection, the station will attract around 40,000 passengers.

○ **项目概况**

历时两年施工，耗资9 000万欧元的奥地利格拉茨Europaplatz广场交通枢纽在预算内按时完工。设计师对车站前面的区域进行了重新设计，新建了一个凸出于站房的、被当地人称为"黄金眼"的屋顶。这个屋顶成为整个广场的中心。曾经最为现代化的有轨电车现在行驶于地下，有4条线路与车站直接相连。由于当地交通枢纽提供国际运输服务，总站因而每年将接纳大约4万名乘客。

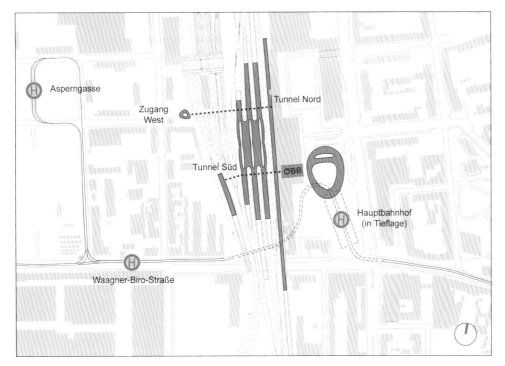

The project has indeed had consequences for development all around the station site. Making the large plaza more attractive has proven to be just the first project for the area. The area's improvement has reached a second stage with the redesign of Annenstraße, where construction work is ongoing.

Site Plan 区位图

项目确实带动了车站周边地区的发展。事实证明。使这一巨大的广场更有吸引力只是该地区发展规划的第一步。作为第二阶段发展计划的 Annenstraße 大街改造目前正在实施。

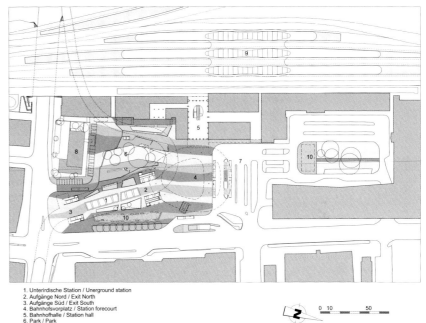

1. Unterirdische Station / Unerground station
2. Aufgänge Nord / Exit North
3. Aufgänge Süd / Exit South
4. Bahnhofsvorplatz / Station forecourt
5. Bahnhofshalle / Station hall
6. Park / Park
7. Busbahnhof / Bus terminal
8. Hotel / Hotel
9. Bahnsteige / Platforms
10. Fahrräder / Bicycles

Layout 布局图

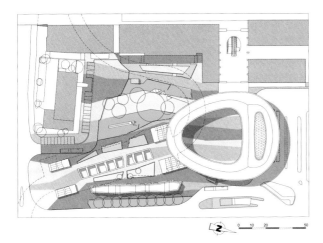

Site Plan 总平面图

◯ Design Highlights : The Ring Roof Called "Golden Eye"

The new double-length stop for the tram lines 1, 3, 6 and 7 is situated underground below the green space at Europlatz. The stop is open to the sky above the lines, with covered waiting areas. This allows natural light to illuminate the platforms and provides ventilation that, in the event of a fire, is capable of clearing smoke from the station without need of additional systems. Passengers can also see the neighbouring hotel and the sky, which helps them to orientate themselves. Escalators and elevators provide disabled access at plaza level.

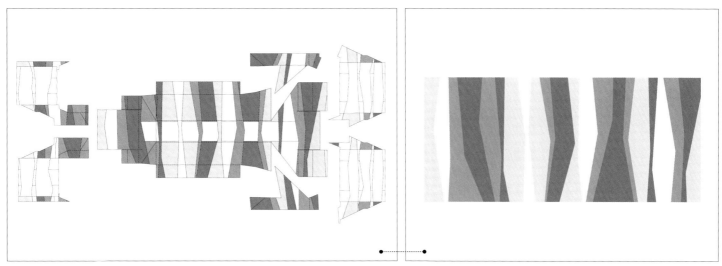

Section 剖面图

○ 设计亮点："黄金眼"环形屋顶

1、3、6、7号电车线路的新站比原站长一倍，位于Europlatz广场的绿地的下方。轨道上方是露天的，候车区域没有遮蔽物，以便为此空间引入自然采光和通风，而且在遭遇火灾时无需启用额外的系统便能让烟雾快速散去。因为能够看到邻近的酒店和天空，乘客在这里也不会迷失方向。自动扶梯和电梯能够帮助残障人士到达广场。

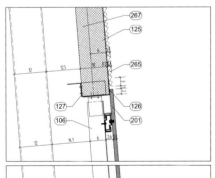

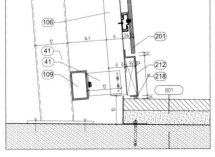

The wall and ceiling surfaces of the station are covered with a stripe pattern of fibre-reinforced concrete panels in three different grey tones. The angled and structured surface makes the station seem to have been 'cut' from the ground, with these cuts extending onto the surface of the plaza.

车站墙面的屋顶由 3 种不同灰度的增强纤维混凝土板覆盖着，它们所组成的斜向条纹图案使车站看起来像从地面"切割"下来一般，这些切口又进一步延伸至广场表面。

Detail 细节图

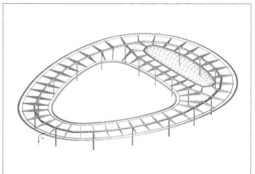

Design Concept 设计概念图

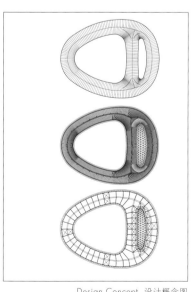

Design Concept 设计概念图

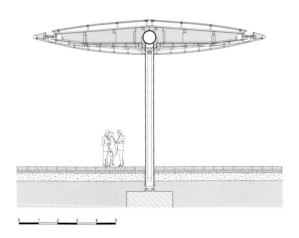

Canopy Section 环形屋顶剖面图

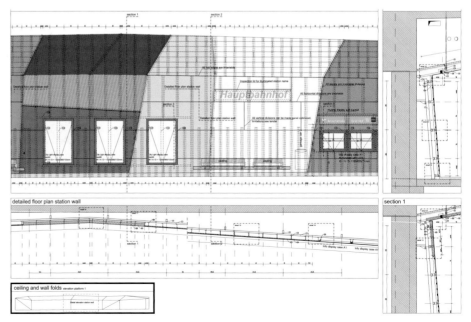

Station Wall Section Detail 车站墙体剖面细节图

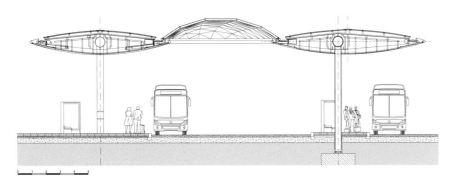

Bus Station Canopy Section 车站顶篷剖面图

The most significant element of the redesign is the covering of the plaza – an oval, ring-shaped disk that provides a ring of shelter to the station concourse and bus stops. The ring roof, called 'Golden Eye' by the locals, creates an 'outer concourse' outside the existing station concourse. Viewed from below, the roof's covering reflects a slightly distorted version of the stripes of the plaza pavement, passengers and vehicles, resembling a movie screen projection of their movements.

重建设计中最重要的元素就是广场上方的椭圆的环形屋顶，它是车站广场和汽车站的遮蔽物。这个被当地人称为"黄金眼"的环形屋顶在现有车站大厅外面创建一个"室外大厅"。从地下看，屋顶反射着广场路面以及乘客和车辆稍微变形的影像，如同一个巨大的电影屏幕。

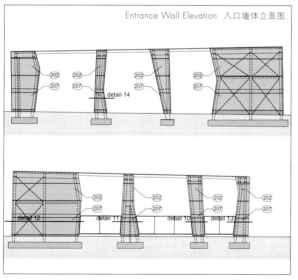

Entrance Wall Elevation 入口墙体立面图

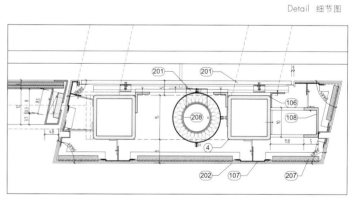

Detail 细节图

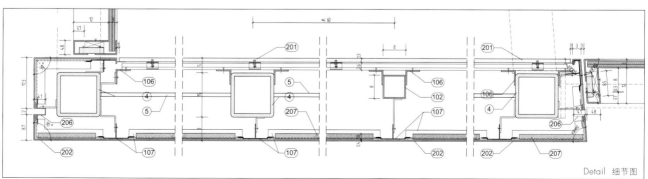

Detail 细节图

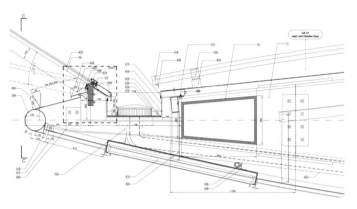

Edge of the Roof Section 屋顶边缘剖面图

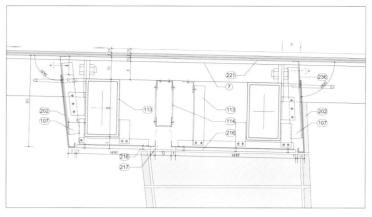
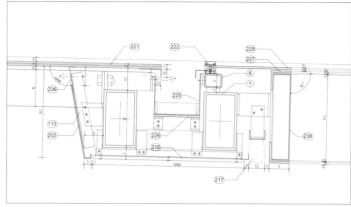
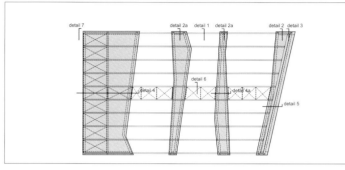
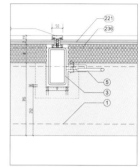
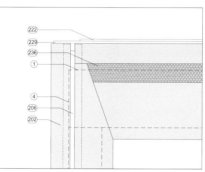

Detail 细节图

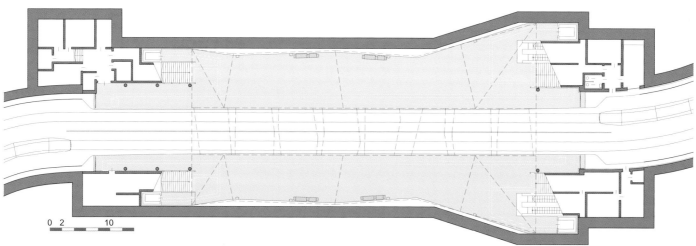

Platform Plan 站台平面图

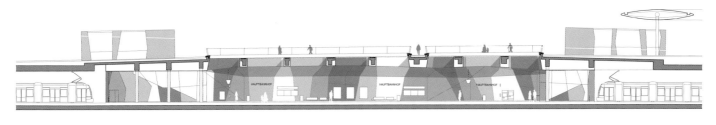

Station Longitudinal Section 车站纵向剖面图

◐ Design Details

The primary steel construction of the approx. 3,000 m² projecting roof is made up of articulated circular-tube supports, the spine, which connects the individual supports, and radial ribs. The total tonnage is approx. 420 tons. Prefabrication on site was used throughout construction. The limited space available at the construction site and short time window available for erecting the structure had also to be taken into consideration. The solution selected was bolting the entire structure together, without on-site welding.

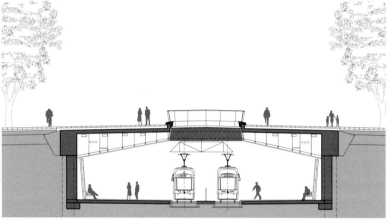

Section 剖面图

○ 设计细节

大约 3 000 m² 的钢结构屋顶由铰接式的圆管支柱,连接支柱的脊柱状物体以及呈放射状的像肋骨一样的物体构成。整体重量约 420 吨,整个施工工程采用了现场预制的方式。此外,我们不得不考虑施工现场有限的空间以及架设结构的短暂工期。最终的解决方案是通过螺栓连接整个结构,不采用现场焊接的方式。

The construction of the new below-ground stop offered the opportunity of adapting the heterogeneous look of the station plaza and the unsatisfactory layout of routes.

建设新的地下车站为我们提供了对车站广场风格不一致的外观和令人不满的路线布局进行调整的机会。

The station's stripe pattern extends over the paving of the plaza area, and shades into differently designed green zones. The polygon beds of plants, designed by landscape architects 3zu0, and a rolling lawn topography not only makes relaxing in the station plaza much more visually attractive, but also provides various recreational uses.

地下车站的条纹图案延伸至广场路面，转变成不同的绿地。由 3zu0 景观事务所设计的多边花坛以及起伏的草坪不仅为车站广场营造了休闲的气氛，而且提供了多种不同的休闲用途。

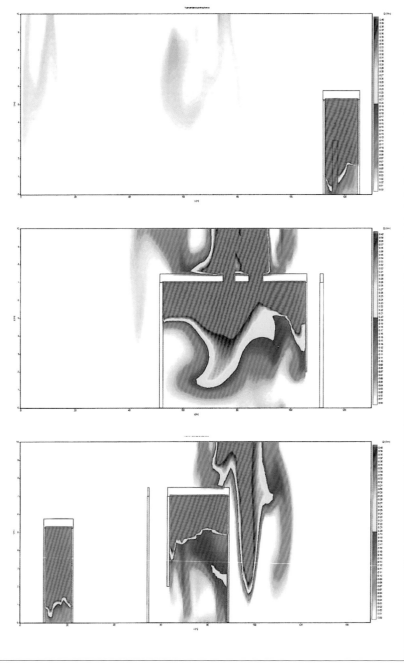

Design Concept Plan 设计概念图

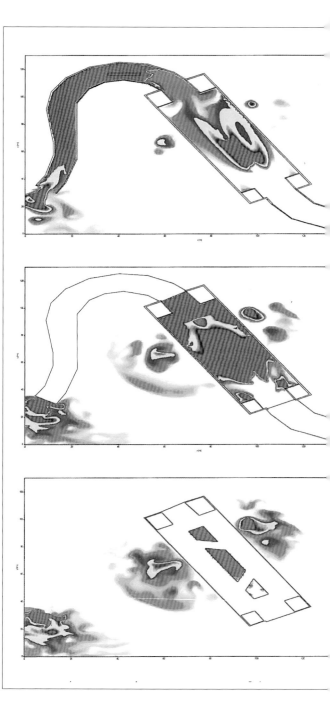

Design Concept Plan 设计概念图

The improvement of bike lane access to the front of the station and creation of new covered bike storage was also part of the building project.

此外，该项目还包括通往车站前的自行车通道以及带遮顶的自行车库。

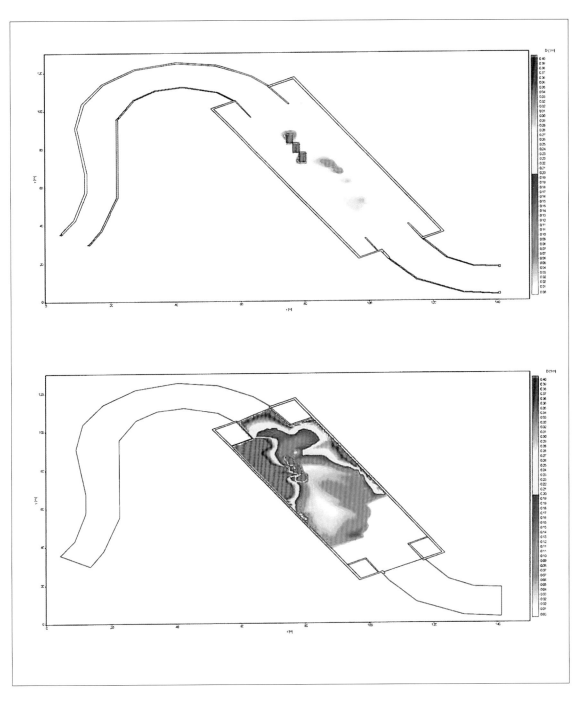

Design Concept 设计概念图

- **Architects:** BIG
- **Project Leader:** David Zahle
- **Team Members:** John Pries Jensen, Henrik Kania, Ariel Joy Norback Wallner, Rasmus Pedersen, Annette Jensen, Dennis Rasmussen
- **Partner in Charge:** Bjarke Ingels, David Zahle
- **Client:** HELSINGØR MUNICIPALITY, HELSINGØR MARITIME MUSEUM
- **Location:** HELSINGØR, DK
- **Area:** 6,500 m²

- 设计公司：BIG
- 项目负责人：David Zahle
- 设计团队：John Pries Jensen、Henrik Kania、Ariel Joy Norback Wallner、Rasmus Pedersen、Annette Jensen、Dennis Rasmussen
- 合作伙伴：Bjarke Ingels、David Zahle
- 客户：赫尔辛格市，赫尔辛格海事博物馆
- 项目地点：丹麦赫尔辛格
- 面积：6 500 m²

丹麦国家海事博物馆

Danish National Martime Museum

◯ **Project Overview**

The Danish Maritime Museum had to find its place in a unique historic and spatial context; between one of Denmark's most important and famous buildings Kronborg Castle and a new, ambitious cultural centre. This is the context in which the museum has proven itself with an understanding of the character of the region and especially the Kronborg Castle.

◯ 项目概况

丹麦国家海事博物馆处于独特的历史环境和空间环境之中——位于丹麦最重要和最著名的建筑 Kronborg Castle 与新建的气势恢宏的文化中心之间。在这样的环境下，项目在充分认知本区域特别是 Kronborg Castle 的建筑特点后，彰显自我风格。

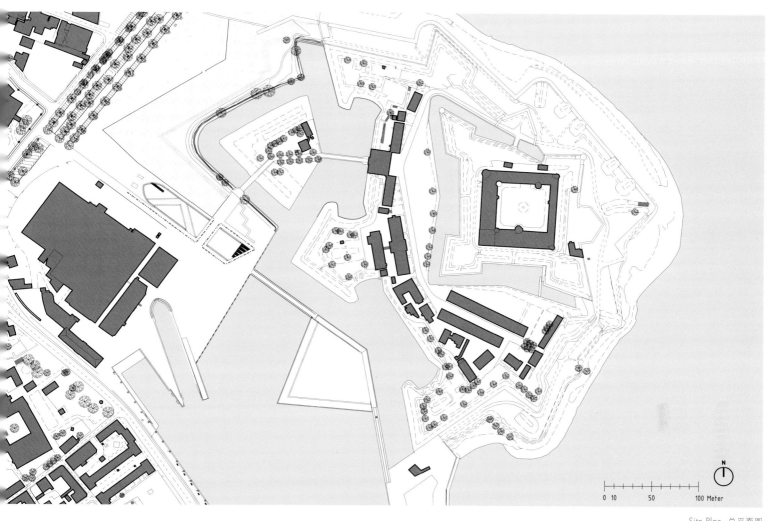

Site Plan 总平面图

◯ **Design Highlights :The Bridge Become the Short-cut for the Visitors**

The Danish Maritime Museum is like a subterranean museum in a dry dock. Leaving the 60 year old dock walls untouched, the galleries are placed below ground and arranged in a continuous loop around the dry dock walls - making the dock the centerpiece of the exhibition - an open, outdoor area where visitors experience the scale of ship building. A series of three double-level bridges span the dry dock, serving both as an urban connection, as well as providing visitors with short-cuts to different sections of the museum. The harbor bridge closes off the dock while serving as harbor promenade; the museum's auditorium serves as a bridge connecting the adjacent Culture Yard with the Kronborg Castle; and the sloping zig-zag bridge navigates visitors to the main entrance. This bridge unites the old and new as the visitors descend into the museum space.

overlooking the majestic surroundings above and below ground. The long and noble history of the Danish Maritime unfolds in a continuous motion within and around the dock, 7 meters below the ground. All floors - connecting exhibition spaces with the auditorium, classroom, offices, café and the dock floor within the museum - slope gently creating exciting and sculptural spaces.

设计亮点：连接桥成为出入捷径

丹麦海事博物馆就像是干船坞中的一个地下博物馆。建筑师没有对有着六十年历史的船坞墙做任何变动，而是将画廊设在了地下，围绕干船坞墙的画廊被设置成连续的环状，形成一个开放的室外区域，游客可以在这里体验到船舶建造的规模。三座双层的连接桥横跨

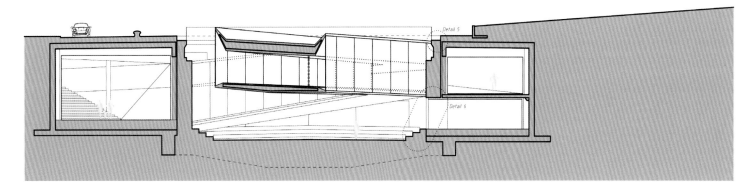

Section 剖面图

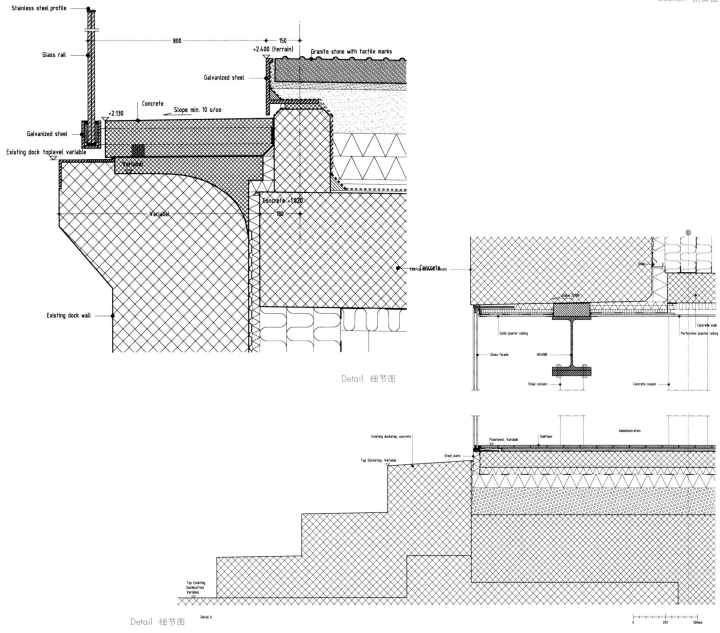

Detail 细节图

Detail 细节图

在干船坞上，既发挥着连接城市的作用，也为游客参观博物馆的不同部分提供了捷径。海港大桥封闭着船坞的同时也成为了一个海滨长廊；博物馆礼堂变成了一座连接着相邻的文化庭院与 Kronborg 城堡的桥；倾斜的"之"字形桥引导游客来到主入口。这座桥将新与旧连在一起，游客向下进入博物馆空间时将体验到新与旧的融合，在这里可以看到地上、地下宏伟的景观。

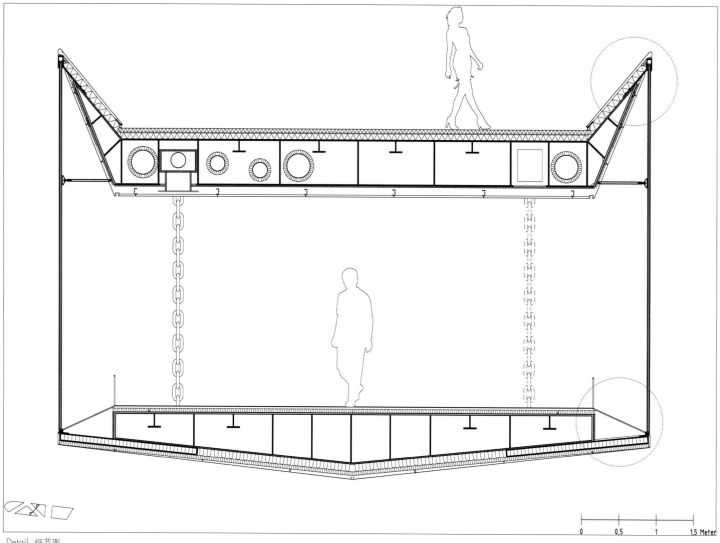

Detail 细节图

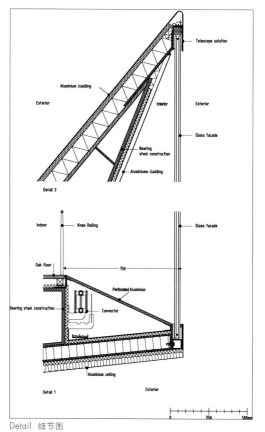

Detail 细节图

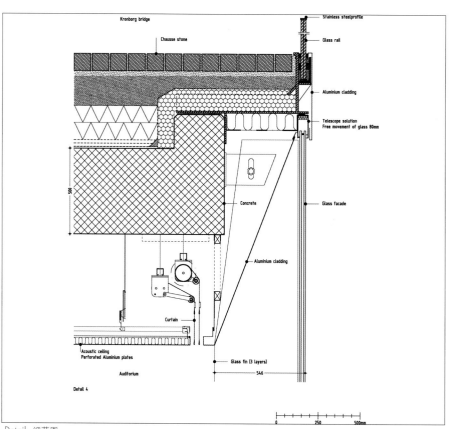

Detail 细节图

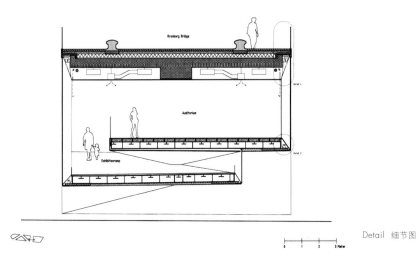

Detail 细节图

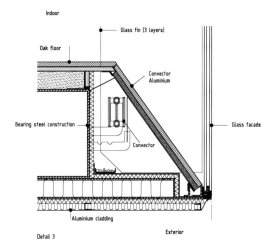

Detail 细节图

Detail 3

○ **Design Details**

The long and noble history of the Danish Maritime unfolds in a continuous motion within and around the dock, 7 meters below the ground. All floors - connecting exhibition spaces with the auditorium, classroom, offices, café and the dock floor within the museum - slope gently creating exciting and sculptural spaces.

○ 设计细节

丹麦海事光辉悠久的历史连续地展示在位于地下 7 m 的船坞内及其周围。将展览空间与礼堂、教室、办公室、咖啡厅和博物馆内船坞楼连在一起的所有楼层都呈现出缓坡的形式，创造出令人兴奋的、极具雕塑感的空间。

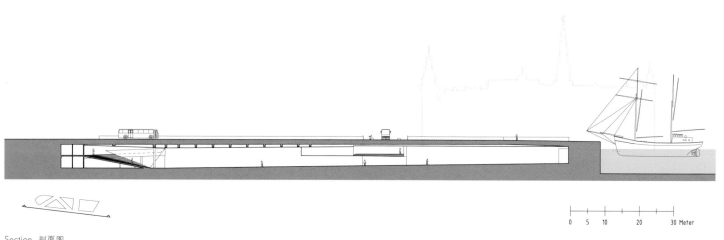

Section 剖面图

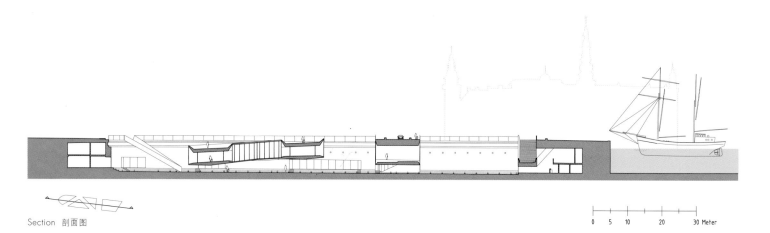

Section 剖面图

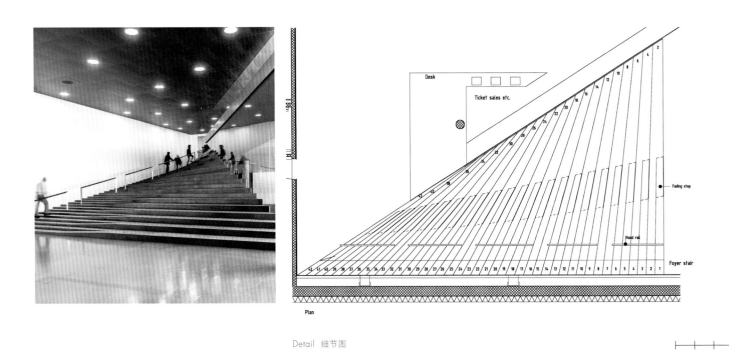

Detail 细节图

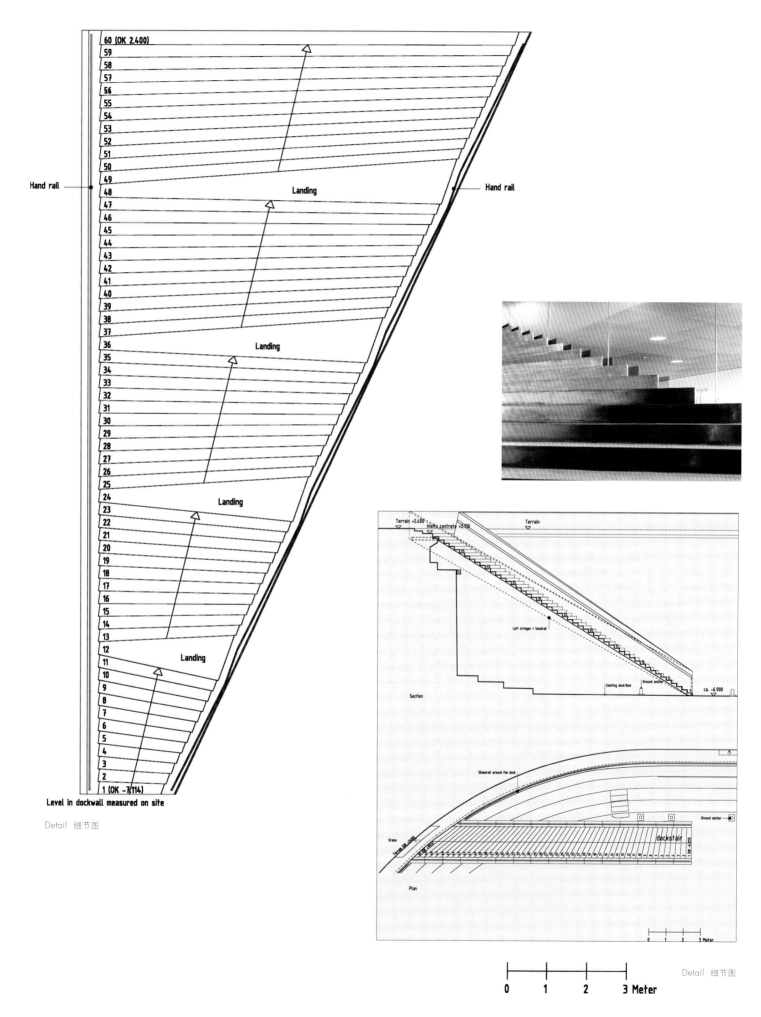

Detail 细节图

Detail 细节图

- **Architects:** C. F. Møller Architects
- **Engineer:** Peter Korsbæk
- **Collaborators:** Balslev
- **Client:** Vendsyssel Museum of Art
- **Location:** Hjørring, Denmark
- **Size:** 3,800 m²

- 设计公司：C. F. Møller Architects
- 工程师：Peter Korsbæk
- 合作者：Balslev
- 客户：Vendsyssel Museum of Art
- 地点：丹麦海宁
- 面积：3 800 m²

海宁 Vendsyssel 艺术博物馆

Vendsyssel Museum of Art

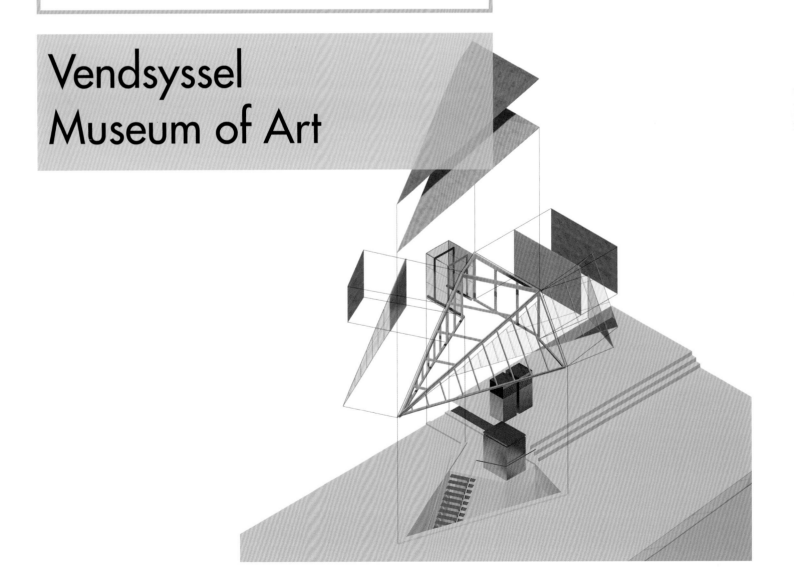

◯ **Project Overview**

Vendsyssel Museum of Art is housed in the converted former Bech's clothing factory, built at the end of the nineteenth century. The conversion has transformed the highly detailed industrial building into a simple and straightforward cultural building which unites the style of the past with a contemporary expression.

◯ 项目概况

Vendsyssel 艺术博物馆坐落于前身是 Bech 服装厂的地方,是在 19 世纪末建成的。经过改建后,那座非常精致、复杂的工业建筑变成了一座简单明了的、把以前的风格和现代的表现方式结合在一起的文化建筑。

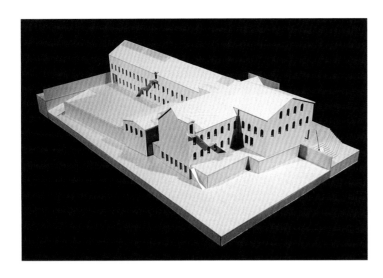

With a larger exhibition area and more storage facilities, it has become possible for the first time to exhibit the permanent collection (primarily of north Jutland artists after 1945) and temporary exhibitions at the same time. The new facilities also enable the museum to live up to modern requirements towards lighting, indoor climate and security.

因为改建，博物馆获得了较大的展览空间和更多的储藏设施，所以它能够同时展览永久性藏品（主要是1945年之后，北日德兰半岛艺术家们的藏品）和临时性的藏品。新的设施，使博物馆在采光、室内环境与照明方面更能满足现代的要求。

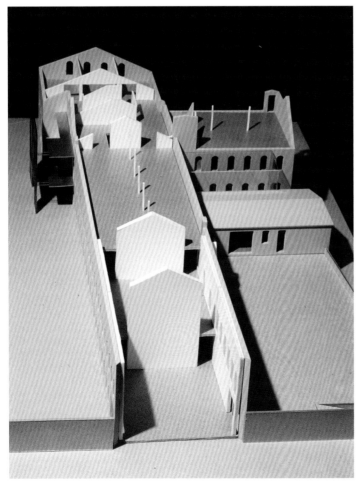

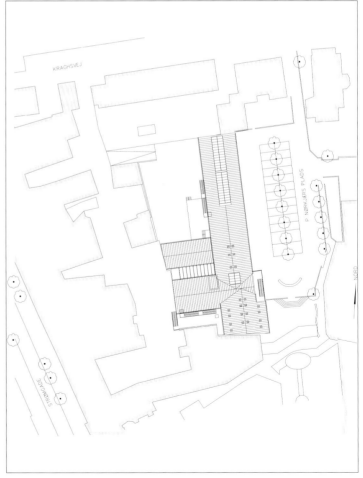

Site Plan 总平面图

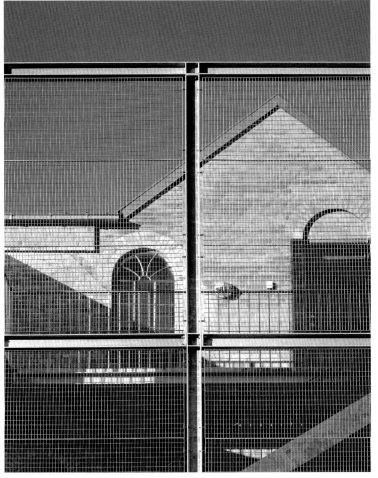

○ Design Highlights : The Screen is Used As a Lighting Element and an Entrance

The textile factory lies in the centre of the town, and the changes here emphasised the need to make the entrance to the new museum more visible. This has been accomplished mainly by establishing a large glass and steel façade screen, which is visible from all areas of the square. The screen can be used as a lighting element as well as to provide support for signage.

○ 设计亮点：屏幕作为入口和照明使用

位于镇中心的纺织厂和这里发生的改变，决定了新博物馆需要一个更为显眼的入口。这主要通过在外面建造一个大型的玻璃钢屏幕来完成，这个屏幕无论从广场的哪个角度都可以看到。屏幕可以作为照明元件使用，同时解决了树立博物馆标志的问题。

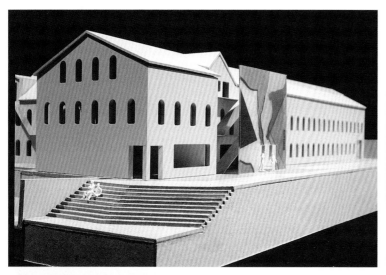

The museum consists of a long main building and two smaller side buildings, connected at right angles. The largest side building houses a library, an Internet cafe and a cinema, while the smaller building has been converted into a guest apartment and workshop atelier for visiting artists. In the glass-roofed space between the two side buildings there is a sculpture garden, and the new museum also offers teaching facilities, workshops, photo labs and photo workshops.

博物馆由一个长长的主建筑和旁边两座小建筑组成。它们连成直角。最大的侧边建筑里设立一个图书馆、一个网吧和一家电影院，同时，较小的侧边建筑改建为一个客房公寓以及一个提供给来访艺术家的工作室。在玻璃屋顶，两座侧边建筑之间的空间，有一个雕塑花园。这个新博物馆同样提供教学设施、工作室、照片处理实验室和照片处理工作室。

The main building houses the foyer and exhibition spaces, with the foyer serving as the connecting point for the museum's activities. The exhibition spaces are organized with the temporary exhibitions on the ground floor, to allow rapid changes and access to the outdoors, while the permanent exhibitions are on the first floor, and the darker graphic art rooms are located in the attic.

设计师在主建筑设立门厅和展览空间,并把门厅作为博物馆活动的连接点。展览空间位于一楼,用于组织临时性展览活动,以便可以应付突如其来的变化,并可以顺利地连接到室外。而永久性展览通常在二楼举行,颜色较深的图片艺术展览室则设立在阁楼。

In order to ensure sufficient wall space to hang the artworks, new interior partitions have been established which cover every second window aperture in the main building's façade. In front of the other windows are matt glass panels to diffuse the incoming daylight.

为了保证有足够的墙面悬挂艺术品,新的室内分区已经建立了,这可以使主建筑的里面有足够的窗孔用于悬挂艺术品。在其他窗户的前面有玛特玻璃板,用于扩散照进来的日光。

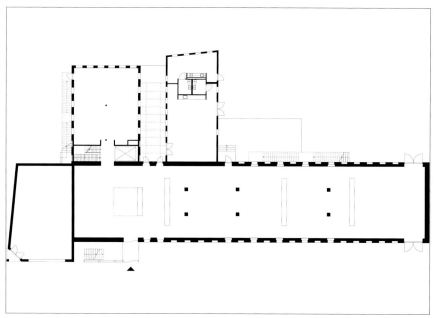
Ground Floor Plan 一层平面图

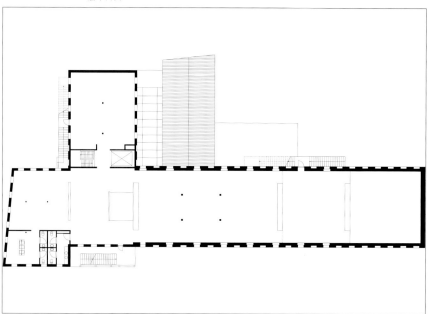
First Floor Plan 二层平面图

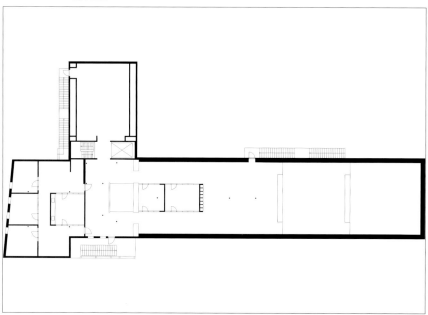
Second Floor Plan 三层平面图

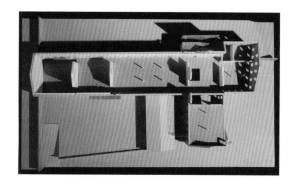

◐ Design Details

The later Vendsyssel Museum extension is placed one level below ground in the western part of the museum courtyard, and consists of a clean and simple exhibition space housing a new knowledge centre for one of Denmark's most important 20th-century painters, Niels Larsen Stevns. The artist has from the beginning had a prominent place in the collections of the museum, which contain art relating to the Vendsyssel area.

○ 设计细节

Vendsyssel 博物馆在 2008 年到 2009 年作了一次扩建。扩建工程在博物馆西院的地下一层进行。扩建的空间包括一个干净、简约的展览空间。这个展览空间设立在为丹麦 20 世纪最重要的画家之———Niels Larsen Stevns 而设的一个新的知识中心。这个艺术家的作品从一开始就在这个包含着 Vendsyssel 地区艺术特色的博物馆占据着重要的地位。

The design of the new centre is deliberately subordinate and architecturally sensitive, by creating the extension to the very rigorous, aesthetic and authentic museum building

as an underground addition. Access to the new exhibition space is via a glass prism in the courtyard, with direct access to the existing

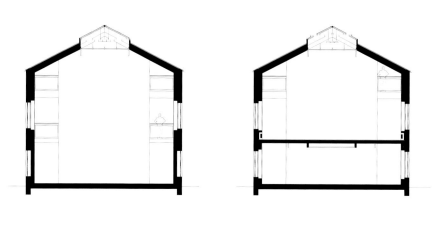

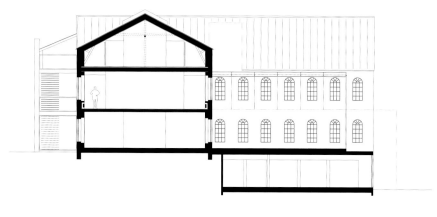

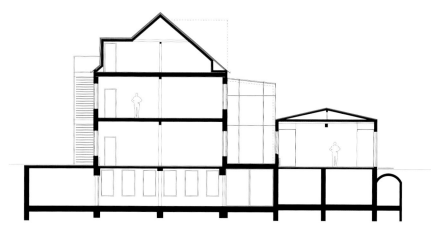

Section 剖面图

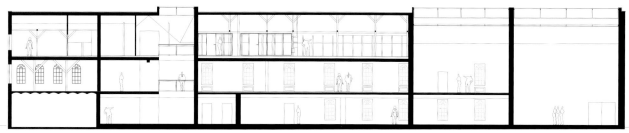

Section 剖面图

exhibition rooms on the ground floor. The glass prism and elongated skylight stretching along one side of the new exhibition space is a key element, drawing light into the space and giving a sense of space and sky above. These elements also create a natural connection between the extension and the existing museum.

通过为这个建造非常严格的、非常注重审美的和正宗的博物馆扩建一个地下附属空间，新的展览中心的设计理念显现出来了。设计师把这样一个重要的展览中心定位为从属空间，其目的是为了在建筑设计上引起人们的注意。人们可以通过院子里的一个带有玻璃棱镜进入新的展览中心，这个入口也能直接连通一楼的展览厅。玻璃棱镜和细长的天窗沿着新展览中心一侧延伸，是能顺利扩建博物馆的一个关键因素。另一因素是让光线照进展览空间，使下面的人们能看到上面天空。以上这些因素也为博物馆和其底下的空间创建一个自然的连接。

- Architects: ON-A
- Client: ON-A
- Photographer: Lluis Ros
- Location: Barcelona, Spain
- Floor Area: 400m²

- 设计公司：ON-A
- 客户：ON-A
- 摄影师：Lluis Ros
- 地点：西班牙巴塞罗那
- 占地面积：400m²

ON-A 建筑事务所新办公室

ON-A'S New Office

◯ **Project Overview**

The new ON-A location arises from the need to expand as a laboratory of architectural theory and practice, and gives the designers some rooms for research projects related to new digital technologies and production process parameters.

◯ 项目概况

建立 ON-A 建筑设计公司新的办公室是因为该公司需要扩建一个实验室,用于进行建筑理论的研究和实践,以及为设计师提供一些用于研究新的数字技术和产品工艺参数的办公室。

◯ **Design Highlights :Unique Entrance**

The new office is an underground office with a green glass door and a strict electronic guard system. The light on the green grass door create a soft and unique feeling for the office.

◯ **设计亮点：富有个性的入口**

新办公室建在地下，入口装上绿色的玻璃门，并设有严格的电子门卫系统。当办公室开灯以后，灯光照在绿色的玻璃门上，显得柔和而富有个性。

Design Details

The office has only one layer, but it is divided into several rooms according to the different function. The office contains conference hall, research offices and a special area for exhibition. The new space is characterized by open, collaborative research and production solutions conceived to generate the solutions demanded by the collective groups within the field of design and architecture. The office, with exhibition capacity, allows for interdisciplinary connections associated with the evolutionary process of our changing society.

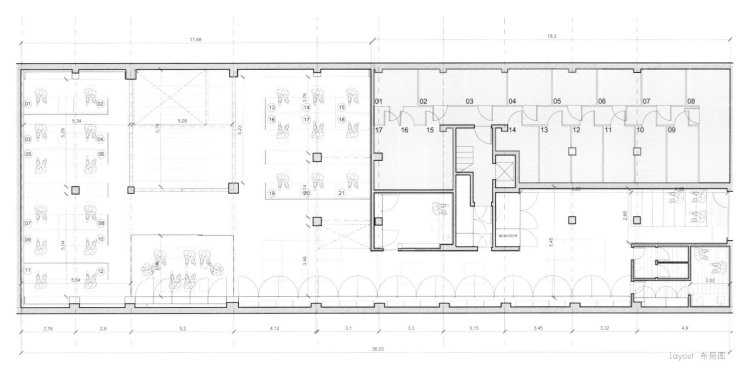

Layout 布局图

○ 设计细节

办公室虽然只有一层,但却包含个几个功能不同的房间,设有会议室、研发办公室和特定的展览区。这个办公室的特点是开放,设计师可以在这里共同研究,提出设计和建筑方面的设想,为团队解决问题。作为一个新的文化中心,办公室还具有举办展览的能力,让设计师根据日新月异的社会环境进行跨学科的研究。

The height and width of the office is designed so suitably that it can provide an easy space for the member of the office. The main color of the office is white, which makes the interior environment clear and bright. A square patio is built in the middle of the office which makes plenty of nature light inside. Therefore, the room is in a good lighting condition.

办公室楼梯和楼层的高度及宽度设计非常合适。员工在这个面积不大的空间里不会感到拥挤。为了让室内环境显得更整洁、明亮,办公室选用了白色为主色调。设计师在办公室的中央设置了一个方形的天井。白天阳光可以通过天井照进室内,使室内显得更明亮。这大大改善了办公室的采光情况。

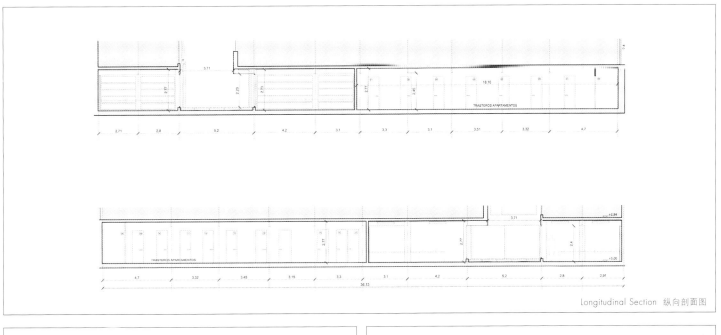

Longitudinal Section 纵向剖面图

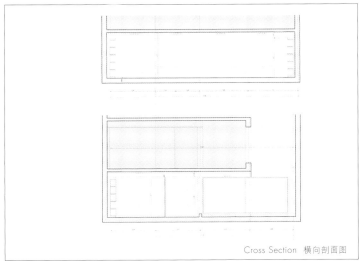

Cross Section 横向剖面图

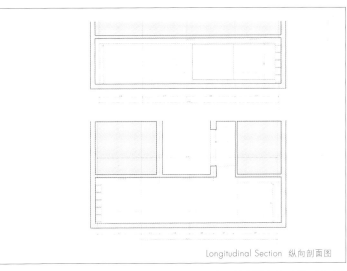

Longitudinal Section 纵向剖面图

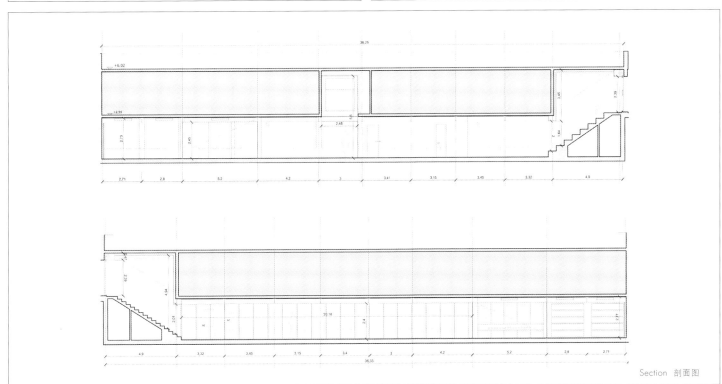

Section 剖面图

- **Architects:** Peter Ruge Architekten
- **Location:** Sofia, Bulgaria
- **Site Area:** 12,390m²
- **Construction Floor Area:** 2,430m²

- 设计公司：Peter Ruge Architekten
- 地点：保加利亚索菲亚
- 总面积：12 390m²
- 建筑面积：2 430m²

Station 20 地铁站

◯ **Project Overview**

Acting as a primary metro connection between Sofia's city centre and the airport, and the hub of a major commercial redevelopment plan, the design of Station 20 site seeks to establish clear pathways, form active connections and harmonise architecture and landscape.

◯ **项目概况**

作为索菲亚市中心和机场的一个主要的地下连接和主要的商业重建计划的中心,Station 20地铁站设计在这样的一个位置上,是为了在活跃的连接点之间建立明确的通道、协调当地的建筑和景观。

Site Plan 总平面图

◯ Design Highlights : Optimised Connection Can Afford Large Quantities of the Persons

Main entry to Station 20's underground concourse level is accessed via a vast corner plaza, providing a public space for circulation between the station services and the existing bus service infrastructure. The sweeping canopy of the entrance hall emerges out of the landscape as a wave, simultaneously pulling the structure up as it pushes the plaza down into the ground to meet with the level of the concourse. Finely cast steel elements with a non-flammable canvas lining stretched underneath, create an elegant structure and a glowing interior, achieved through concealed and integrated lighting.

◯ 设计亮点：经优化的连接适应高客流量

人们可以通过广阔的角形广场，来到 Station 20 的主入口，来到地下的大堂。在入口处，有一个服务站，它改善了地铁服务和现有的公交车基础服务设施之间的循环运作。在入口大厅已被清扫的顶篷上所显现的像波浪一样的景观，既把整个线性结构往上拉，也把广场推到地下，使其与大堂保持在同一水平线上。设计师使用精钢制造的元件和不易燃的帆布向下拉伸，并通过隐蔽的集成照明系统，创造出一个优雅的结构和一个明亮的室内空间。

In response to the flow of passengers with the quantity of 120,000 persons, the stations concourse level has been extended to accommodate the maximum passenger loads expected, thus reducing congestion through passenger control systems. Visual connection from all entrances into the halls have been optimised to promote passenger safety, assist station security and maximise natural light.

针对每天120 000人次的客流量,地铁站大堂所处的层面已作了延伸,以适应人们预期中能负载的最大客流量,以此调控客流运输系统,减少拥塞。从所有入口到大厅的连接已被优化,以起到更好地保证乘客的安全,协助站内治安的作用,让地铁站获取最多的光照。

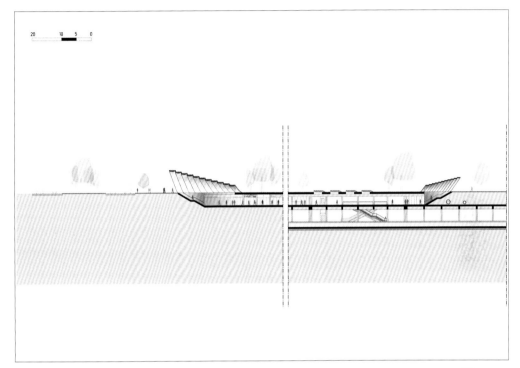

Section 剖面图

The concourse is further extended to include a shopping court, capitalising on high passenger flows, whilst providing future potential for an underpass connection from the station to the planned adjacent shopping centre development.

设计师对大堂作进一步扩展,并扩建了一个购物广场,利用高客流量,为地铁站和相邻的购物中心开发的连接通道提供更大的发展潜力。

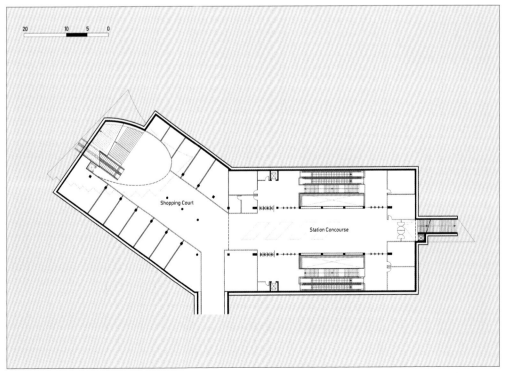

Plan 平面图

◯ Design Details

The sites linear structure is inspired by the pattern and scale of the neighbouring high-density housing blocks of the Druzhba residential quarter. The angular direction of the landscape reflects the natural structure of the site boundary, generating a visual language that informs the division of public spaces and creates pathways, guiding pedestrian flow to and from the station.

○ 设计细节

地铁站采用线性结构,这一灵感来自于邻近的 Druzhba 住宅小区的高密度住宅群的模式和规模。从景观的角度来看,这样的设计体现出该地方边界的自然结构,形成了一种视觉语言,向人们展示了公共区域的划分,并显示了新建的路径,引导着进出地铁站的人流。

Chapter Four
第 4 章

Landscape design above ground and in the entrance can be an efficient way to create a unique city view. Generally speaking, there are three methods. Firstly, we design landscape of the top of underground architecture and entrance to reflect harmoniously as a beautiful view on the ground square. Secondly, we use different landscape forms to design the top of the underground architecture and entrance to feature the part above ground. Thirdly, we use green plants to hide the part above ground and entrance. This method is suitable for the underground architecture with large volum and exceed the projection line range. In this chapter, all the classic works will provide you with a pleasant journey.

The Space with Natural Landscape and Man -made Marvels

自然造景的空间

在地下建筑的顶部和出入口造景，能形成一种独特的城市景观。一般来说，造景的方法有以下三种：第一，合理地布置地下建筑地上部分的位置，使顶部和出入口以景观的形式出现在广场上；第二，室外广场上突出地面的地下建筑顶部和出入口，设计成独立式景观；第三，通过绿色植物来隐藏地下建筑，尽量将建筑顶部和出入口布置在绿化带里或者其他较隐蔽的地方，这种方法适用于当地下建筑的范围较大，远远超出地面建筑投影线范围时的情况。在本章中，经典项目将带你游历赏心悦目的旅程。

* **Architects:** *BIG*
* **Project leader:** *Joao Albuquerque, Gabrielle Nadeau*
* **Team members:** *Maren Allen, David Tao, Salvador Palanca, Marcos Bano, Lucian Racovitan, Ryohei Koike, Camille Crépin, Elisa Wienecke, Léna Rigal, Paolo Venturella, Tiina Liisa Juuti, Jeff Mikolajewski²*
* **Client:** *Groupe Auchan*
* **Location:** *Paris, Frech*
* **Size:** *800,000 m²*

* 设计公司：*BIG 建筑事务所*
* 项目负责人：*Joao Albuquerque、Gabrielle Nadeau*
* 团队成员：*Maren Allen、David Tao、Salvador Palanca、Marcos Bano、Lucian Racovitan、Ryohei Koike、Camille Crépin、Elisa Wienecke、Léna Rigal, Paolo Venturella、Tiina Liisa Juuti、Jeff Mikolajewski²*
* 客户：*Groupe Auchan*
* 地点：*法国巴黎*
* 面积：*800 000 m²*

Europa City
活动中心

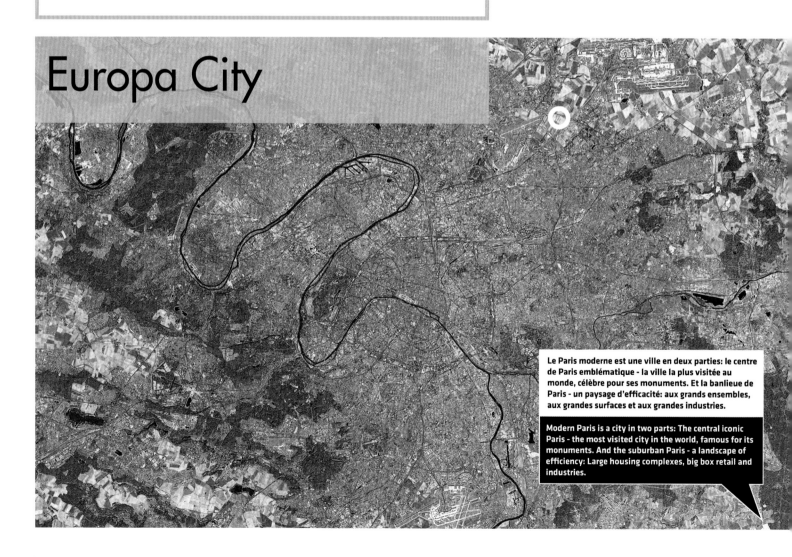

Le Paris moderne est une ville en deux parties: le centre de Paris emblématique - la ville la plus visitée au monde, célèbre pour ses monuments. Et la banlieue de Paris - un paysage d'efficacité: aux grands ensembles, aux grandes surfaces et aux grandes industries.

Modern Paris is a city in two parts: The central iconic Paris - the most visited city in the world, famous for its monuments. And the suburban Paris - a landscape of efficiency: Large housing complexes, big box retail and industries.

○ Project Overview

Europa City will offer, on an unprecedented scale, a mix of retail, culture and leisure around a defining theme: Europe, its diversity, its urban experiences and its cultures. Rather than orienting Europa City towards the highway, the designers integrate it as the natural center of a new business district.

○ 项目概况

Europa City 有着空前的规模，其内部围绕一个明确主题划分出零售、文化和休闲区域。Europa City 展示了城市多种多样的生活方式、城市带给人们的体验以及城市的文化底蕴。设计师没有把欧洲城设置在正对着高速公路的地方，而是把它整合起来，建造成具有自然生态元素的新商业中心。

Aerial View 鸟瞰图

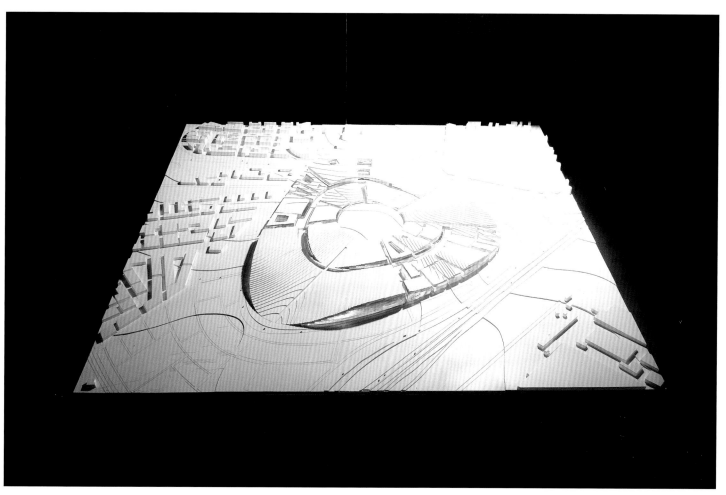

Design Concept 设计概念图

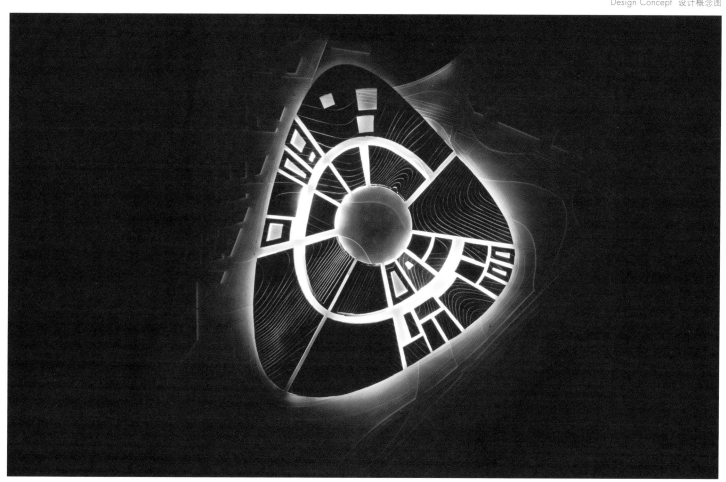

Design Concept 设计概念图

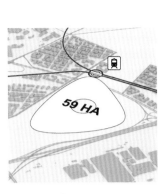

RER AND METRO STATION
The footprint of 80 ha dedicated to Europa City is in direct connection with the metro / RER station, which will be fully integrated into the urban design of Europa City. Its eastern facade also provides a good visibility of the complex from the A1.

Design Concept 设计概念图

THE PEDESTRIAN BOULEVARD
The metro and RER station is the starting point of a pedestrian street "Tour d'Europe", which will be the backbone of Europa City. The shops will be concentrated in its periphery, recreating for the visitor the stimulating experience of an open medieval commercial street.

Design Concept 设计概念图

MAXIMUM HEIGHT
The footprint is then extruded to the maximum height permitted by the take-off cones of Le Bourget airport, in order to provide sufficient space for functions that require a high clearance height.

Design Concept 设计概念图

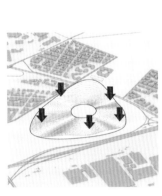

VALLEYS AND PEAKS
Where the program does not require a high clearance height, the roof is selectively lowered, allowing access to the roof from the pedestrian promenade. These were also designed to frame some remarkable views towards Paris.

Design Concept 设计概念图

CONNECTIONS TO THE SURROUNDINGS
Side streets are drawn perpendicular to the pedestrian promenade, meeting at the center of Europa City to provide shortcuts for visitors, and radiating outward to create connections to the neighborhood.

Design Concept 设计概念图

GREEN CONTINUITY
Like the development plan originally proposed for the "Coeur du Triangle", a central park secures a connection between the "Carré vert" and "Buttes Saint-Simon", looping the belt of suburban parks between Roissy and Le Bourget. The Central Park is a public facility not exclusive to the visitors of Europa City, but for all the neighboring communities.

Design Concept 设计概念图

Design Concept 设计概念图

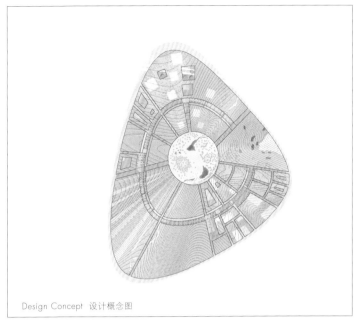

Design Concept 设计概念图

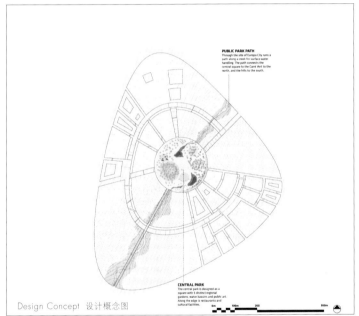

Design Concept 设计概念图

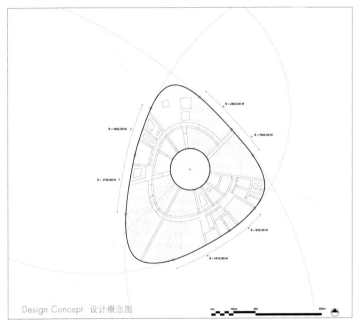

Design Concept 设计概念图

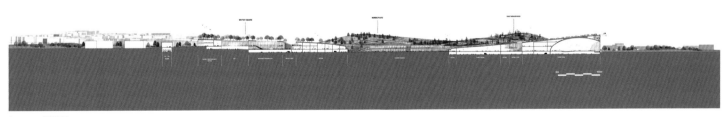

Section 剖面图

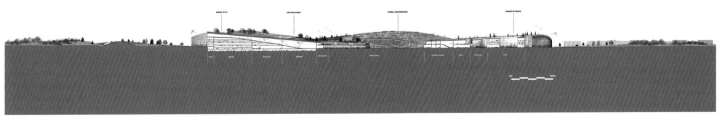

Section 剖面图

○ Design Highlights : Improve the Development of Surrounding Architecure

The designers propose an urban form that combines dense city with open landscape, exploring the urban and green potentials of the site at once. They propose a commercial city that becomes a gathering point for the surrounding neighborhoods– a new kind of commercial center that is blended into the urban texture, and includes rather than excludes the surrounding city.

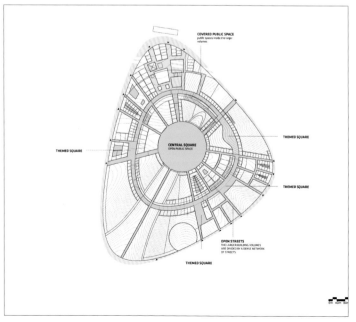

Plan 平面图

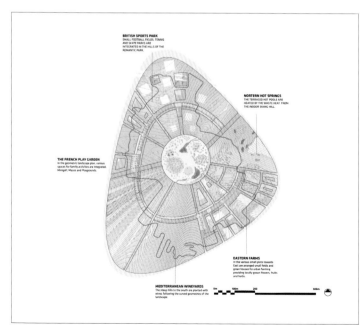

Plan 平面图

设计亮点：带动周边建筑的发展

设计师提出一个新的城市形态方案，那就是既保证活动中心有足够的人流量，又保证不影响周围的生态景观，让城市和绿色生态因素能更好地结合在一起。在设计师的构想中，这个商业中心将成为了一个人口聚居点，以带动周围的建筑。它将融入城市的生活圈中，与周围的城市区域连成一体，不与周边的区域产生冲突。

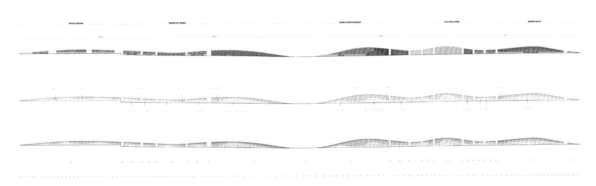

Elevation 立面图

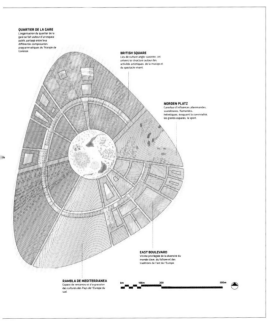

Plan 平面图

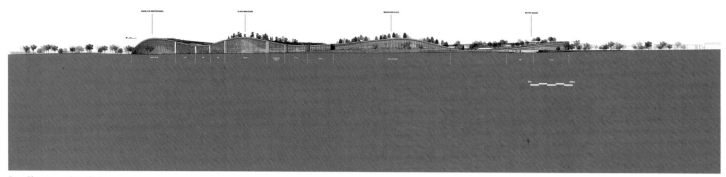

East Elevation 东立面图

North Elevation 北立面图

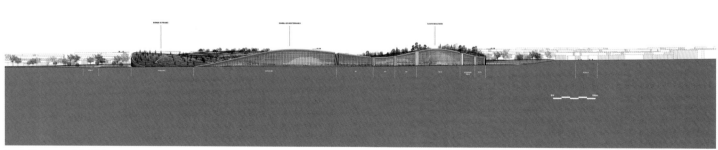

South Elevation 南立面图

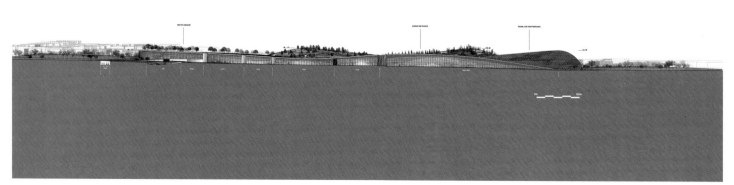

West Elevation 西立面图

The programs of Europa City are organized along an internal boulevard with a mix of retail, entertainment and cultural programs on both sides. The boulevard forms a continuous loop travelling through six different areas each celebrating a different lifestyle. Along the boulevard, public bicycles and electric public transport bring visitors quickly around. The circular street creates at the same time surprising spatial experiences and a clear overview - It allows you to get lost, and still find your way.

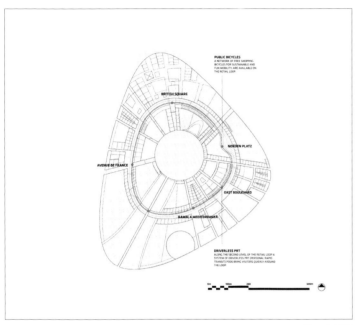

Plan 平面图

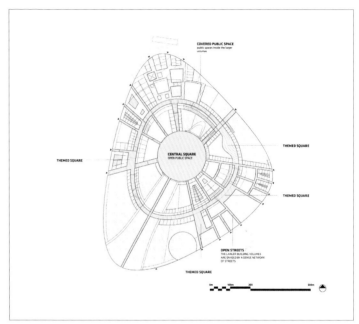

Plan 平面图

设计师沿着内侧的林荫大道设计 Europa City 项目，该项目将在路的两侧设置零售，娱乐和文化区域。大道看上去像一个连续不断的圆环，通过六个不同的区域展示不同的生活方式。游客可以在这里通过骑公共自行车和电动的公共交通工具快速地转上一圈。环形的街道能同时带给人们不同寻常的空间体验和清晰的整体景观。哪怕你迷路了，也能找回原来的位置。

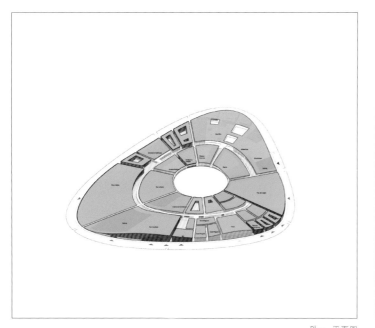
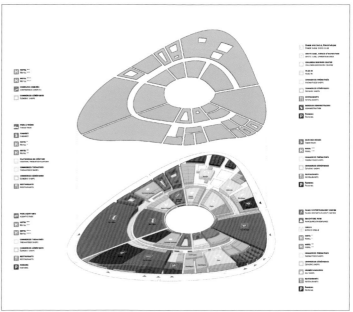

Plan 平面图

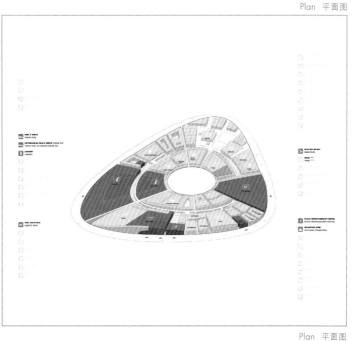
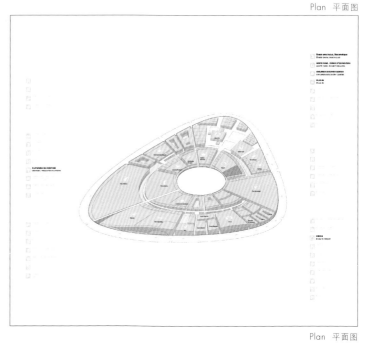

Plan 平面图

◯ Design Details

The customers can come to the underground space by the escalator or the spiral stairs. Being similar to ground floor, there are some commercial spaces in the first basement. The first basement is divided into three spaces. Two of them are separated spaces. The other space and the space above are regarded as a large space. One of the separated spaces is enclosed. Some shops will be built here. They form a commercial street. Another separated space is wide and open. Customers can see the commercial facilities on the ground floor and the ceiling of the Europa City here. Some large glass panels also set in the roof in order that there is a good lighting condition for the underground space. The second basement is divided into several spaces, including three car parks, three storage rooms and several spaces for retail. Thanks to the division, it strengthens the connection between the ground floor and the underground space. The underground spaces of different functions play an important role in supporting the facilities on the ground. It is more convenient for customers to visit the Europa City and the condition of logistics is better than before.

○ 设计细节

进入 Europa City 的游客可通过手扶电梯和螺旋式的楼梯来到地下层。跟地上层一样，地下一层包含一些商业设施。设计师根据建筑的形状设立两个独立的空间和一个跟一层相通的空间。其中一个独立空间是相对封闭的，这里将建造一些小商店，形成独特的地下商业街。而另一个独立空间非常宽阔和开放，乘客抬起头能看见一楼的商场和天花板。为了保证地下空间的采光，屋顶的一些地方镶嵌了玻璃板。地下二层被分隔为多个空间，其中包含三个停车场，三个仓储室和多个用于零售业的空间。这种划分使地下与地上的联系更加紧密。不同的地下空间为楼上的各种设施提供了强大的支撑作用，有利于疏导人流和物流。

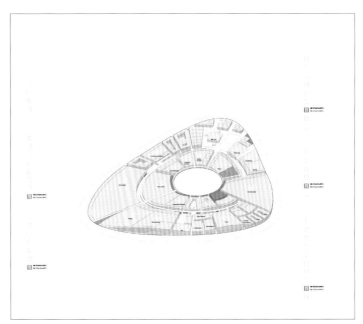

Plan 平面图

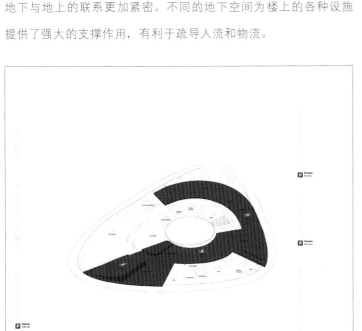

Plan 平面图

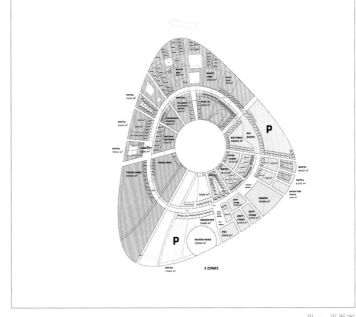

Plan 平面图

- **Architects:** *Spektrum Arkitekter*
- **Location:** *Reykjavik, Iceland*
- **Site Area:** 8,000 m²

- 设计公司：*Spektrum* 建筑事务所
- 地点：冰岛雷克雅未克
- 面积：8 000 m²

雷克雅未克 Ingolftorg 广场

Ingolftorg, Reykjavik

◯ **Project Overview**

Reykjavik's story begins along Aðalstræti which constitute the city's origin and historical center. The designers want this proposal to strengthen Ingolftorg and Vikurgadour as the center of the city , both historical , social and cultural ways . With this proposal, they have pursued to combine the two sites and the new hotel in an overall narrative and design, in addition, to meet the desired performance requirements they create unity and coherence in the town and bring the city a large and prominent square framed by the existing buildings .

项目概况

雷克雅未克的历史起源于 Aðalstræti，一个城市发源地和历史中心。设计师希望加强 Ingolftorg 和 Vikurgadour 这两个地区的建设，使它们成为城市的中心。带着这一目的，他们力求把这两个地方和一家新的酒店结合起来，形成一个整体的方案和设计思路。此外，为了满足所需的建筑设施要求，他们加强城镇的统一性和连贯性，并根据现有建筑的架构，在城市中创建一个突出的大广场。

Site Plan 总平面图　　　　　　　　　　　Site Plan 总平面图

○ **Design Highlights :The Square Shape Shows the Staging of Nature's Powerful Forces**

The designers have created an open and lively meeting place, signaling renewal through its contemporary and surprising design, a node crossed by pedestrians, prams and cyclists and it is also a place that invites to relax through time in the sun, play and exercise.

○ **设计亮点：外观体现城市地貌特色**

设计师创建一个开放的、活跃的会议场所，通过现代的、优秀的设计重建信号，行人、推婴儿车的人和骑自行车的人交错经过这里，人们可以在这里享受阳光，尽情玩乐，锻炼身体，以此来放松身心。

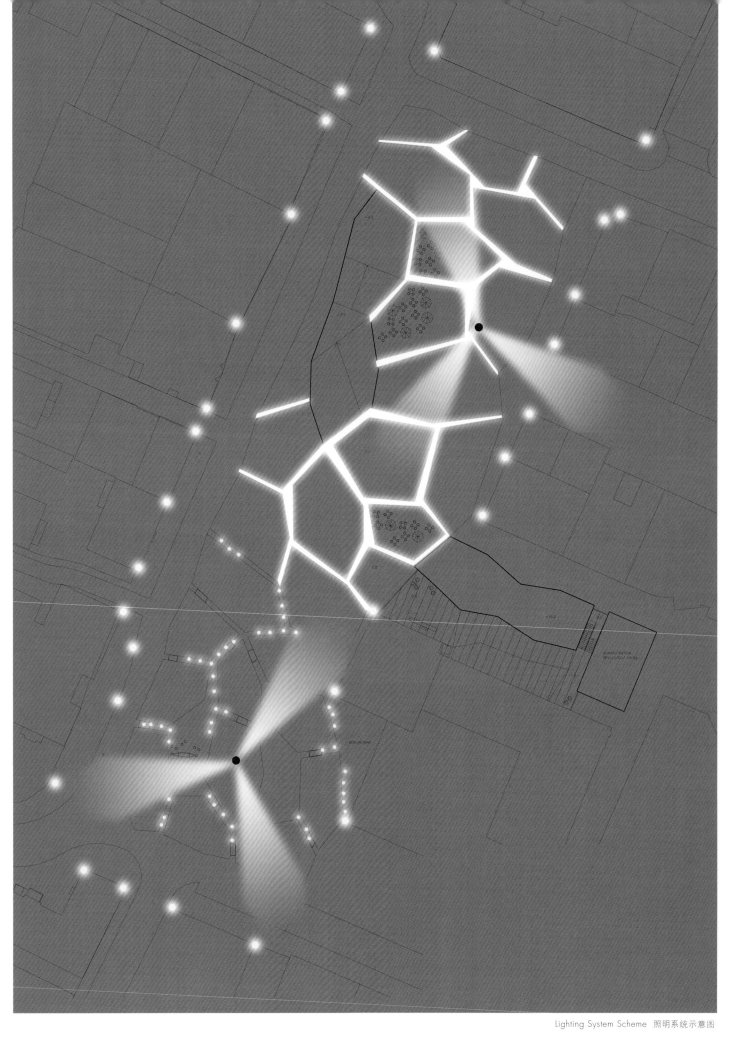

Lighting System Scheme 照明系统示意图

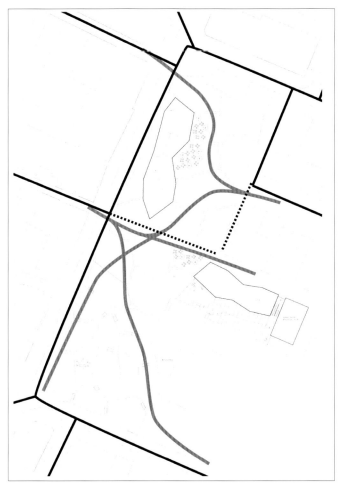

Kinetonema Scheme 动线示意图

Based on Iceland's unique nature, northern lights, the volcanic and geothermal activity and subsurface tension and mystery, they have created an "under and above" ground artificial landscape with a strong identity. Chunks of ice basalt formations, the moving continental plates and earthquakes, have been the inspiration for the estate and the square shape which appears as an urban interpretation and staging of nature's powerful forces. Therefore appears as a cracked square which may be regarded as cracks or flakes in the crust of ice. In these "scabs" or "flakes" there is a big part of the hotel where is indeed situated.

冰岛有不少独特的自然现象，这里会出现北极光，有不少火山，会出现地热运动和地壳变动，还有其他神秘的自然现象。考虑到这一点，我们在地下和地上都创建了人工湖，这些人工湖在建筑结构上起到重要作用。冰块状的玄武岩地层，不断运动的大陆板块和地震现象，为设计师设计商业业态和广场的形状提供了灵感。设计师就是要让建筑的外观体现出城市地貌的特色，让人们透过这座建筑看到大自然力量的强大。故把建筑设计成地面开裂的广场，让人感觉到地面刚受到很大冲击，导致有很多裂缝，或者地面上都是一片一片的冰块。酒店的其中一个面积较大的部分就建在这布满裂缝、由块状物堆砌而成的地方。

Plan 平面图

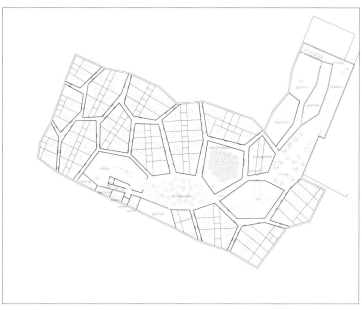

Plan 平面图

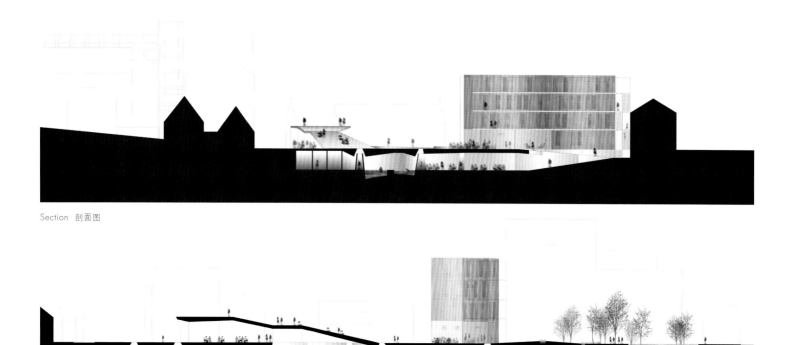

Section 剖面图

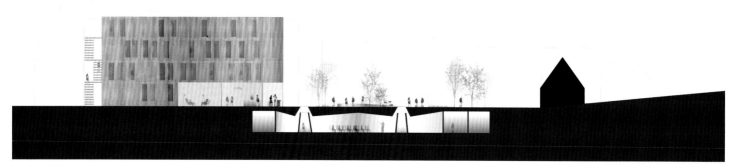

Section 剖面图

Section 剖面图

○ **Design Details**

Placing the hotel underground is mainly proposed to offer Reykjavik city centre a worthy and open square with a certain size, where the area to the extent possible be reserved for the city's public life. In addition, an underground hotel will be a unique hotel with spectacular and beautiful space that can offer an attractive and unique experience. The fusion of building and landscape can be considered a contemporary translation of Iceland's traditional turf houses where the buildings appear buried in the landscape with a heavy layer of soil and roofs.

○ **设计细节**

把酒店建立在地下的一个主要目的是为雷克雅未克市提供一个有价值的、开放的、有一定规模的广场，以便在一定程度上保留城市的公共生活。此外，这个地下酒店会是一个独特的酒店，它有着壮观美丽的景色，能为顾客带来独特的体验，让人流连忘返。把酒店建在景致独特的地面之下，被厚重的土壤和屋顶覆盖着，这种建筑与景观的融合，在人们看来，是从现代的角度，构建冰岛传统的草房子。

- **Architects:** LAVA (Laboratory for Visionary Architecture)
- **Client:** Abu Dubai Future Energy Company
- **Location:** Masdar, UAE
- **Site:** 96,000m² total (Public Plaza 31,200m², 5-Star Hotel 23,265m², Convention 19,766m², Retail 15,500m², Cinema 6,500m²)

- 设计公司：LAVA (Laboratory for Visionary Architecture)
- 客户：Abu Dhabi Future Energy Company STATUS
- 地点：阿拉伯联合酋长国，马斯达
- 面积：总面积96 000m²（其中公共广场31 200m²，五星级酒店23 265m²，会议厅19 766m²，零售商店15 500m²，电影院6 500m²）

马斯达生态广场

Masdar Plaza Centre

○ Project Overview

The future wellbeing of cities around the globe depends on mankind's ability to develop and integrate sustainable technology. Masdar City is the city of the future; positioned at the forefront of integrating sustainable technology into modern architectural design. Rome, Athens, Florence; most great historical cities have had the plaza, forum, or square at their epicentre – where the life, values, ideals, and vision of the population evolved. Equally, the centre of Masdar must be an iconic beacon that attracts global attention to sustainable technology.

Aerial View 鸟瞰图

○ 项目概况

世界各地城市未来的健康发展，取决于人们的开发能力以及完善的可持续发展技术体系。马斯达展示了城市的未来，将可持续发展技术体系与整个当代建筑设计的整合放在首位。罗马、雅典、佛罗伦萨，这些历史悠久的大城市都设有购物广场、论坛或露天广场。那里成为人们几个世纪以来生活和表达思想的地方。同样地，马斯达的中心也要成为一个标志性的地方，以其可持续发展的技术吸引全世界的目光。

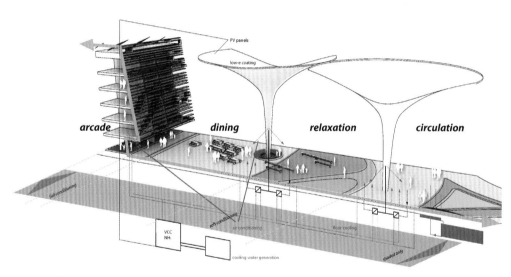

Water Circulation Section 水循环剖面图

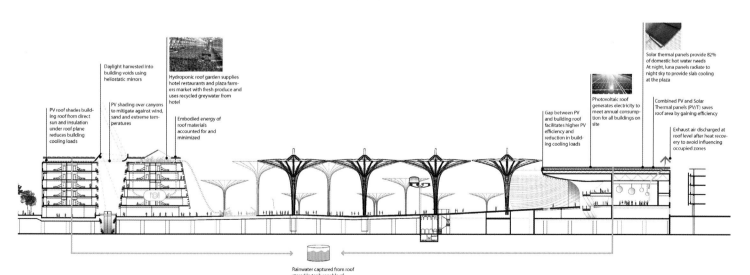

Water Circulation Section 水循环剖面图

Design Highlights: The Open Oasis of the Future

The designers see Masdar Plaza as "The Oasis of the Future": a living, breathing, active, adaptive environ; stimulated by the social interaction of people, and spotlighting the use and benefits of sustainable technology. Hence, our design proposal focuses on the delivery of three key issues:

(1) performance - to demonstrate the use and benefits of sustainable technology in a modern, dynamic, iconic architectural environment

(2) activation - to activate or operate the sustainable technology in accordance with the functional needs of this environment 24 hours a day and 365 days of the year.

(3) interaction - to encourage and stimulate a social dynamic where the life, values, ideals, and vision of the population of Masdar evolve.

设计亮点：开放的未来绿洲

设计师将马斯达广场视为"未来的绿洲"，即一个非常适合生活、呼吸、活动的环境，一个显示可持续技术利用价值的地方。因此围绕以下三点进行设计：

（1）性能展示——以最新的、动态的、图像化的建筑环境值说明可持续技术的回收利用效果。

（2）激活作用——根据一天24小时和一年365天环境功能的需求激活、运行可持续技术。

（3）相互影响——鼓励和激发马斯达中心广场的活力。

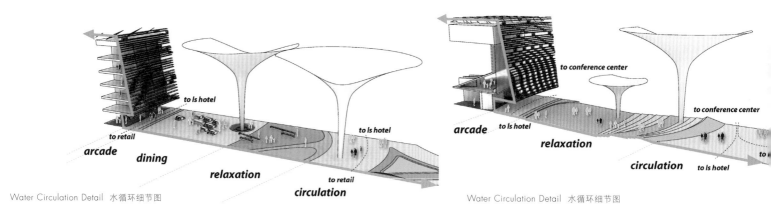

Water Circulation Detail　水循环细节图　　　　　　　　Water Circulation Detail　水循环细节图

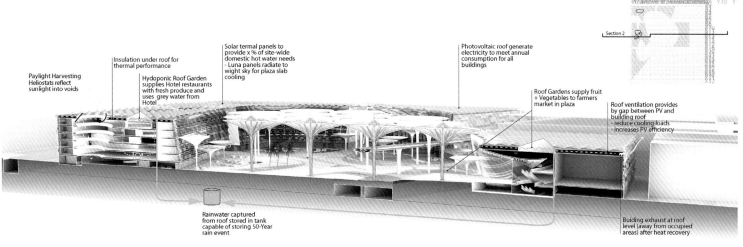

Section 剖面图

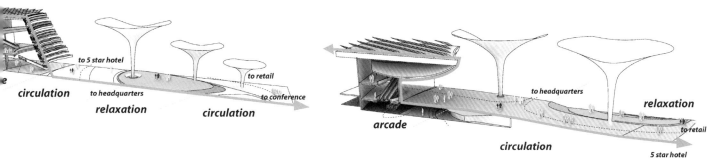

Water Circulation Detail 水循环细节图　　　　Water Circulation Detail 水循环细节图

The "Oasis of the Future" is conceived as an open spatial experience. Whereby, all features; whether hotel, conference, shopping, or leisure, offer the highest quality of indoor and outdoor comfort and interaction. As in the case of an oasis, the Plaza is the social epicentre of Masdar; opening 24-hour access to all public facilities. Giant umbrellas, with a design based on the principles of sunflowers, will provide moveable shade in the day, store heat, then close and release the heat at night in the plaza of a new eco-city in the United Arab Emirates.

"未来的绿洲"被看作是一个开放的、带给人不一样体验的空间。在这里，所有的建筑，如宾馆、会议厅、购物中心或娱乐设施，都为人们带来室内外最高的舒适度和互动性。作为一个绿洲项目，24小时开放的公共设施使广场成为马斯达的社会活动中心。广场上巨大的伞状物可以随意移动，灵活地为人们遮挡阳光和风雨，同时白天储存热量，夜晚将其释放。

There is a unique underground water storage system in the plaza. When it rains, the rainwater is led into the underground water storage space through the pipes. The underground water storage space is so large that it can store 50-year raining event. The establishment of the underground water storage space has a double advantage that both solve the drainage problem and the water recycling problem.

广场上有一个独特的地下蓄水系统。每当下雨的时候，落到屋顶的雨水顺着管道，流到地下的蓄水空间。这个蓄水空间体积非常大，可以储存50年所下的雨量。蓄水空间的设立，既解决了排水的问题，也可以循环利用水资源，可谓一举两得。

The "Oasis of the Future" demonstrates sustainable technology in a user-friendly architectural environment– flexible use of space, outdoor and indoor comfort, and optimum performance.

"未来的绿洲"这个案例表明可持续技术可以以一种更友好的方式植入建筑环境中。在这个过程中，不仅可以灵活利用空间，也可以营造舒适的室内外空间，使整体获得最佳的性能。

- **Architects:** *Gonzalo VAILLO MARTINEZ*
- **Location:** *London, UK*

- 设计师：*Gonzalo VAILLO MARTINEZ*
- 地点：英国伦敦

泰特现代美术博物馆的新地下礼堂
The Cave

The Cave, New Underground Auditorium of the Tate Modern

◯ **Project Overview**

The Cave is located in London. It is an underground auditorium of Tate Modern.

◯ **项目概况**

The Cave 位于伦敦。它是泰特现代美术馆的一个地下礼堂。该建筑是由 Gonzalo Vaíllo Martínez 设计的。

Gonzalo is an architecture student of University of Alcala ETSAG, which is in Spain. He got A for The Cave as the Degree Project. The project of this public building began during the time they have an exchange program with the Bartlett School of Architecture in London.

Gonzalo 是西班牙的高校——阿尔卡拉大学的建筑学学生。他在毕业考试中，凭着这个项目取得了 A 级的好成绩。这个项目是在他们与伦敦的 Bartlett 建筑学校交流的期间开始进行的。

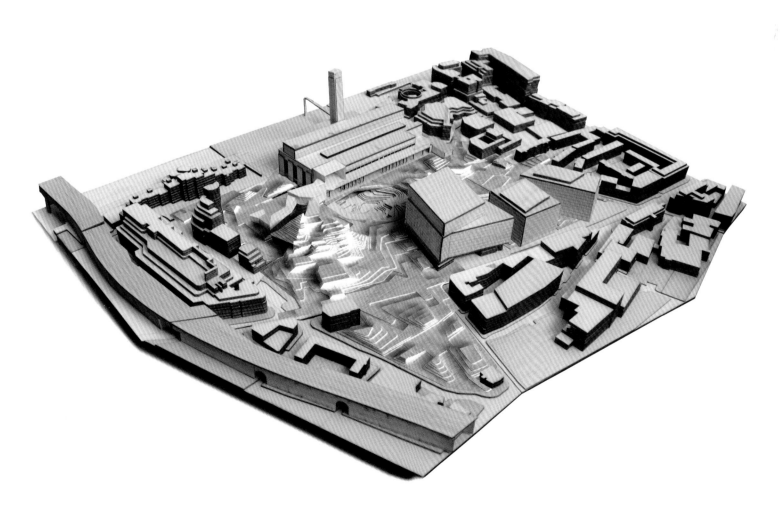

The project is focused on three essential parameters of London: the green areas, the icons and landmarks and the pedestrians. What it is searched is to recover the figure of the pedestrians and involve them again in the city. The public space is the basic network for getting it. Focused on the Tate Modern as an icon which is isolated in its environment (particularly its back side), the project works in two overlapped scales: the necessity to connect the Tate with others landmarks and its own renovation and adaptation for the new social demands. A new park is created whit the intention of solving it and a new auditorium was included. Contemporary art museums constantly need to be by the new social necessities. In this line, they are no longer art containers, but new social centres. Actually we do not know if it is more important visit a new temporary exhibition or drinking a coffee in the top-floor restaurant of the museum where there is a marvelous panoramic view of the city. So, the project includes new different activities to the original Tate's program. A new complete mix-used building is created embedded in the design of a park. This park gives coherence to the whole area and it also connects the museum with the surroundings. the landmark of the Tate's chimney.

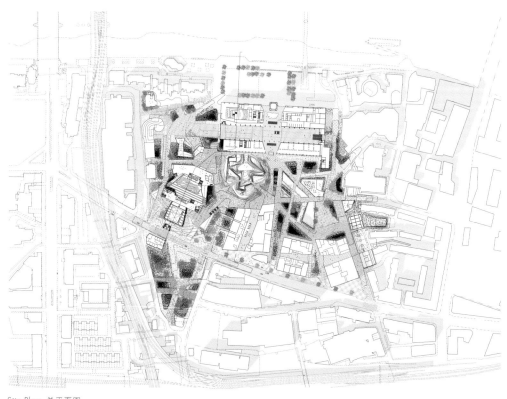

Site Plan 总平面图

项目以伦敦的三个基本参数为重点：绿色区域、图标和地标、行人。设计师要研究的是恢复与行人有关的标志，再次让行人在这个城市畅通无阻地通行。而公共设施为这样的工程的实施提供基本的脉络。作为城市的一个标志，泰特美术馆在城市的环境中显得相对单调和独立（尤其是它的背面部分），它将成为项目中的重点修建对象。项目包含两个相互交叉的部分：第一是把泰特美术馆和其他的已经翻新的地标连接起来；第二是建造适应社会需求的工程。为了解决这些问题，设计师创建了一个新的公园，公园里包含了一个礼堂。在社会必需品日益丰富的今天，建造当代艺术博物馆非常有必要。从这个角度来看，博物馆不再是艺术品的储藏室，而是新的社区中心。实际上，我们不知道参观一个临时展览或者在一个可以看到城市的美丽全景的博物馆上喝咖啡对我们来说，是不是更重要。所以，该项目在原来的泰特美术馆项目的基础上，新设置了各种不同的活动。一座新建的、完整的混合式建筑以嵌入公园的形式来设计。公园连接着整个地区，也同样连接着博物馆和其周边环境，以及泰特博物馆的烟囱这一标志性建筑物。

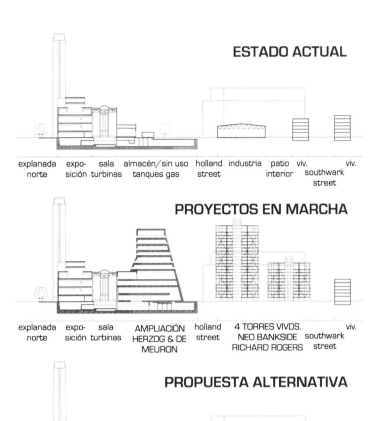

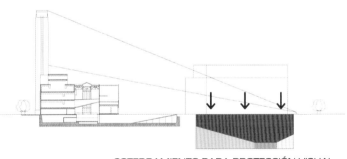

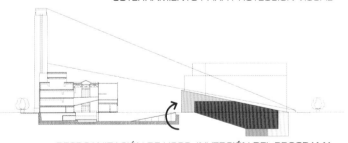

Diagram 示意图

○ **Design Highlights : The Green Roof Becomes a Part of the Park**

Taking into account the importance of the chimney in the skyline of London, the new auditorium is buried for preserving the views over it. This produces an inversion of the normal program of an auditorium, so the accesses and foyers are now over the music halls in directly relationship with the park and its new topography.

○ 设计亮点：绿色屋顶成为公园的一部分

考虑到位于伦敦天际线的烟囱的重要性，新的礼堂建在地下，以保留地上的风景。这产生了一系列与正常的礼堂项目相反的程序，所以，现在礼堂的出入口和门厅设在音乐厅的上面，与公园及其新的地形建立直接的联系。

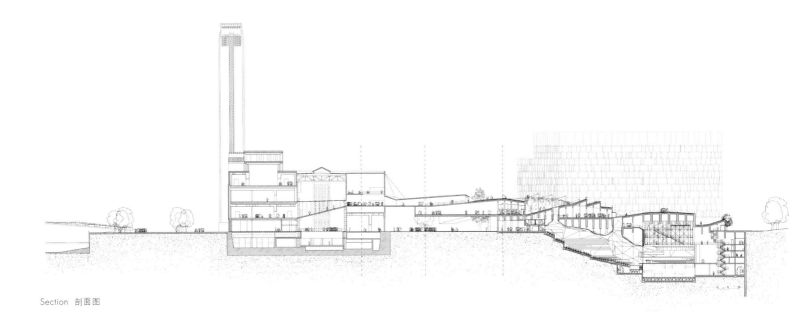

Section 剖面图

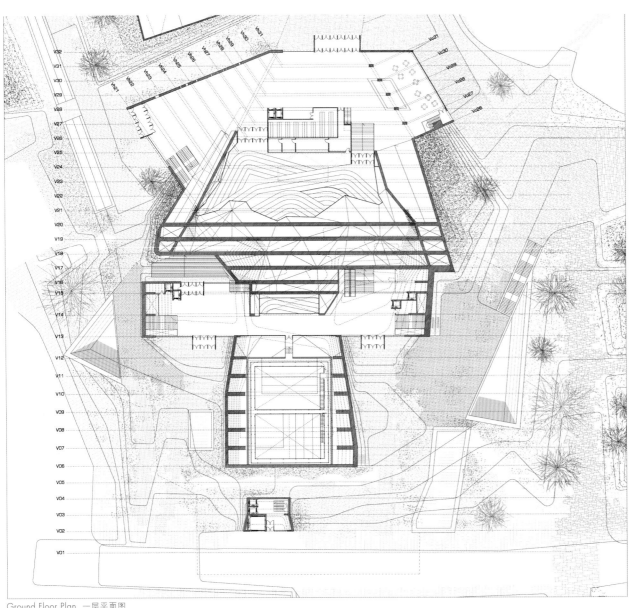

Ground Floor Plan 一层平面图

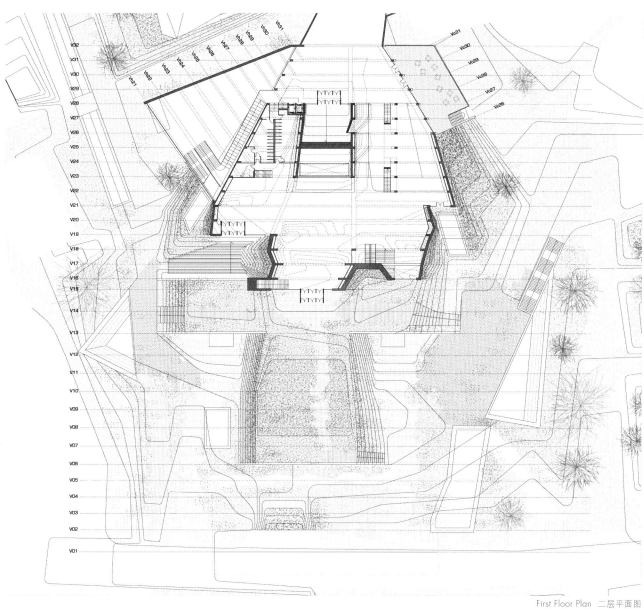

First Floor Plan 二层平面图

The auditorium is made up of three different halls, which have the same stage. This produces that every kind of performance can be showed with simple mechanical partitions depending on the capacity for each event.

礼堂由三个不同的大厅组成,这些大厅在同一水平线上。这让各种各样的表演通过根据不同的功能划分的简单机械分区得以展现。

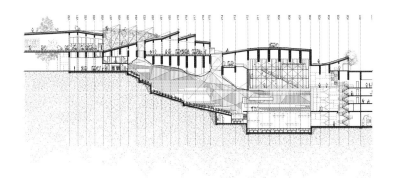

Section 剖面图

Section 剖面图

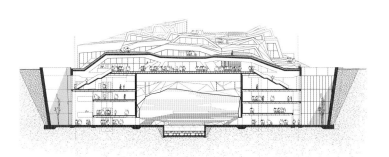

Section 剖面图

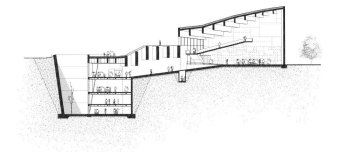

Section 剖面图

The three halls are covered by huge concrete beams. They divide the underground world from the upper public one. Between them the public program of accesses, tickets and multifunctional rooms are developed. Over them, a new topography creates the green roof, which is part of the park. They also create the morphology of the new topography and they are covered by vegetation. Interior halls, foyers and accesses are a mix of this concrete, wood floors and glass walls. The auditorium which is made out of dark wood parquet and an irregular red ceiling of plaster-wood panels provide the good acoustic following the shape of a cave.

三个大厅都被巨大的混凝土梁柱覆盖着。这些梁柱把地下的世界与地上的世界划分开来。在出入口之间，开发了售票处和多功能室。在其上方，设计师根据新的地形，创造了绿色屋顶，这个绿色屋顶将成为公园的一部分。他们同样创造了新的地貌形态，并在上面种植了一片植物。室内大厅，休息室和门厅是用混凝土、木地板和玻璃墙混合建造的。礼堂是用暗色木地板和不规则的、用灰泥木料嵌板建造的红色天花板搭建而成的，它根据洞穴的形状，建造良好的声学系统。

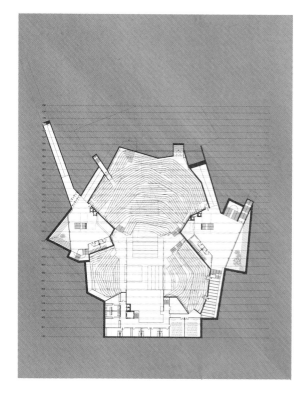

Underground Auditorio Plan 地下礼堂平面图

- **Architects:** *Patkau Architects Inc.*
- **Project Team:** *John Patkau, Patricia Patkau with Thomas Schroeder, Luke Stern, James Eidse, Shane O'Neil, Tyler Brown, Chad Manley, Patric Danielson, Dave Reeves*
- **Client:** *Western Pennsylvania Conservancy*
- **Location:** *Pennsylvania, USA*
- **Total Gross Area:** 523 m²
- **Total Net Area:** 446 m²

- 设计公司：*Patkau Architects Inc.*
- 项目团队：*John Patkau、Patricia Patkau with Thomas Schroeder、Luke Stern、James Eidse、Shane O'Neil、Tyler Brown、Chad Manley、Patric Danielson、Dave Reeves*
- 客户：*Western Pennsylvania Conservancy*
- 地点：美国宾夕法尼亚州
- 总建筑面积：523 m²
- 总实用面积：446 m²

流水别墅上的屋子

Cottages at Fallingwater

◯ **Project Overview**

The meadowland within the native forest above Frank Lloyd Wright's Fallingwater is the site of six new cottages. The meadow, which is a fragment of a recent agricultural landscape, provides a complimentary experience to the surrounding forest from which to appreciate and understand both ecology and cultural history. It opens a clearing in the forest; a clearing that establishes the possibility of a sustainable relationship between human occupancy, site and sun.

◯ 项目概况

在 Frank Lloyd Wright 的流水别墅上面有一片连着原始森林的草地，这里有六间新的屋子。这片草地是近期建造的农业景观的一部分，设计师修整了这片草地，为人们提供一个以欣赏、了解自然生态和文化历史的方式免费体验周围森林的机会。这将在探索森林方面打开一个缺口，让人类的居住能够与阳光、自然环境结合起来，实现可持续发展。

Design Highlights : Cottages from Swells and Falls of Landform

Each cottage forms a small ridge. In series, cottages form swells and falls of landform. Each cottage looks over the low point of the swell in front of it toward the line of White Pine and deciduous forest at the meadow's edge. Cottage-form is an intensification of land-form. Just as Fallingwater is an intensification of the rock outcroppings that characterize Bear Run, the meadow cottages are an intensification of the swelling ground-plane of the meadow, made from the very soil and grasses of the meadow itself.

设计亮点：屋子构成波澜起伏的地形

每个屋子都弯曲成一座小桥的样子。这些屋子构成波澜起伏的地形。在这样的地形中，每间屋子都位于前面向上突起的地面的最低点，沿着与草地边缘连接的白松林和温带落叶林的方向延伸。屋子的形状突出了土地形状的特征，就像流水别墅突出了承受流水冲击的裸露的岩石的特征一样，草上的屋子突出了凸起的接地平面的特征，屋子是由这里的土壤和草制造的。

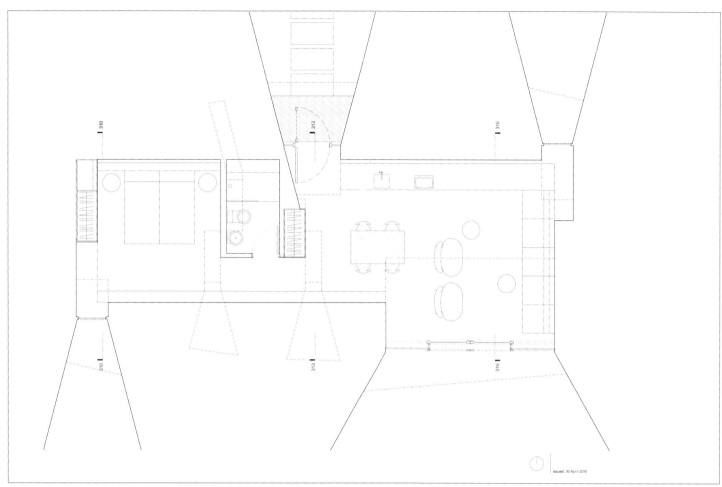
Plan 平面图

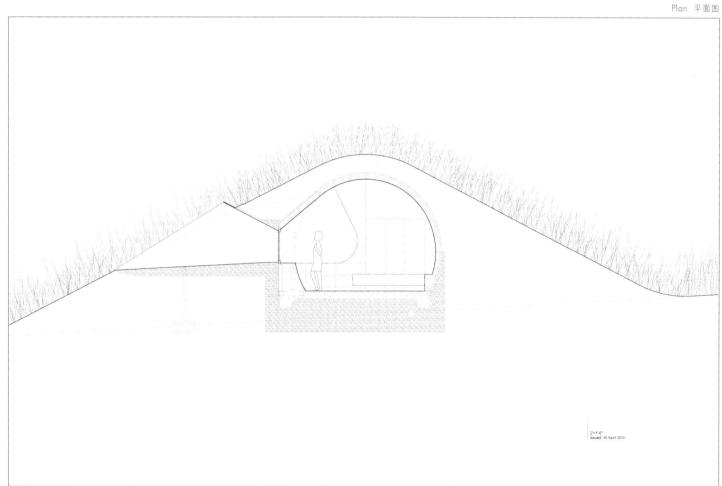
Building Section 建筑剖面图

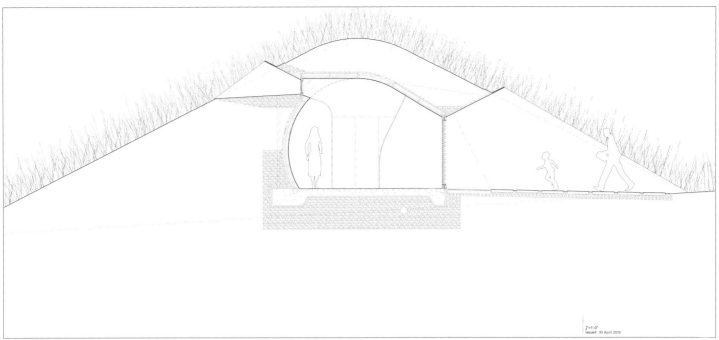

Building Section 建筑剖面图

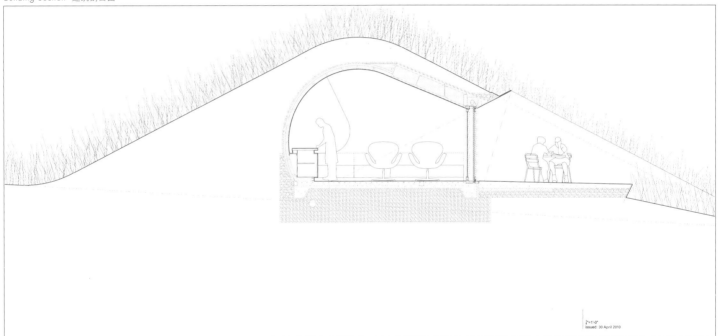

Building Section 建筑剖面图

The cluster of cottages forms a sheltering micro-climate of sunny prospect. Openings in each cottage, cut through the landforms by weathering steel portal frames, connect cottage interiors to the meadowland beyond, silently and passively gathering the sun's energy in the 'shoulder' and in winter seasons, shading openings and providing ventilation during the summer's heat.

这群屋子形成一个遮挡阳光的小环境。每个别墅的开口，通过耐候钢框架连接到草场内外，默默地、被动地用它们的"肩膀"收集太阳能量，它们在冬季遮挡阳光；在炎热的夏季提供良好的通风条件。

The cottages are open in plan, generous with a sense of luminous space. Interiors are surfaced in light toned wood, grounded by concrete floors and sculpted by daylight. A large opening to the meadow focuses the living space, allowing easy access to an outdoor terrace. Bedrooms are more enclosed with double or two single beds as required; extra guests – children - can sleep on the long couch in the living space.

这些屋子在规划上是开放的，能给人带来一种明亮的、宽阔的感觉。室内的墙面是用浅色的木材建造的，地板是用混凝土建造的，造型根据光照情况而设计。一个连接草地的大开口突出了居住的空间，使室内与外面的阳台连接得更顺畅。相比之下，根据要求配置了一张双人床和两张单人床的卧室显得相对封闭一些；如果有额外的来宾——小孩，那就可以安排他们睡在居室的长沙发上。

◯ Design Details

The construction budget for the cottages is very modest so that an extremely cost effective construction strategy is essential. To this end the designers utilizes a standard corrugated steel culvert as the primary building structure and envelope to maximize volume and minimize material and cost. Window and door openings within this envelope are framed with light steel members to create an extremely robust overall enclosure. With the addition of a waterproof membrane to the exterior, and spray applied insulation to the interior, the cottages are fitted out with flexible hardwood veneer plywood interiors.

◯ 设计细节

屋子采用最合算的工程预算，以便让能创造高效益的工程预算策略落到实处。设计师采用一个标准的、刻有波纹的钢制管道，作为建筑的主要结构，让建筑在获取最大的体积的同时，消耗最少的材料，花费最少的资金。在实施这个方案的时候，设计师用轻钢构件制造窗口和门口的框架，以创造一个非常强大的外表面。屋子外部配有额外的防水膜，内部喷涂绝缘涂料，室内采用灵活的硬木胶合板。

Durability and longevity characterize both exterior and interior. The soil and grasses of the siteare a naturally renewable exterior cladding for the primary steel building envelope, weatheringsteel window and door surrounds age to a permanent natural finish, while wood and concrete create a warm low maintenance interior.

屋子的内部和外部都非常坚固，这可以让屋子维持很长的时间。这个地方的土壤和草地都是可再生资源，设计师利用它们熔炼出这些被钢铁包围着的屋子，并且用一些时间风化门窗，在自然而然中打造出一座能永久居住的房子。而木板和混凝土对地势较低的室内环境起到保暖作用。

To further minimize cost, waste and environmental impact, the cottages can be fully or largely constructed within a factory, shipped to the site and attached to a cast-in-place concrete foundation slab and then covered with earth. Key to this approach is the robust structural integrity of the primary steel envelope and an overall unit dimension (not exceeding 4.88 m in width, 4.11 m in height and 25.91 m in length) which can be transported by conventional tractor trailer. To conform to these dimensional limits the living area window extension as well as weathering steel retaining walls would be attached to the primary enclosure after delivery to the site.

为了进一步降低成本，减少浪费，减轻对环境的影响，设计师利用屋子里面的全部或大部分环境建造一个工厂，把材料运到现场，并连接一个现浇混凝土基础底板，然后用泥土覆盖。采用这种方法的关键是保证原钢制造的部分和整个单元的尺寸（宽不超过4.88 m，高不超过4.11 m，长不超过25.91 m）能使房屋结构稳固、完整，并且可以由传统的拖拉机运输材料。要符合这些条件，居住区窗口的扩展要有所限制，同时在材料送到建造场地后，连接主要外墙的耐候钢挡土墙的建造也要控制在一定的尺度内。

Chapter Five
第 5 章

As for the narrow channel with a large number of passengers' flow, it is suitable to design some art drawings with great visual angle and direct visual effect. As for the waiting rooms of underground station, adding some art elements on the details is a great idea to attract passengers when they are waiting for trains. Furthermore, we can reform the space interfaces such as ceilings, grounds and walls according to people's visual feature to provide an open and extended sense for the spaces. For example, underground street can use the turning and reflection effect of the mirror to expand the image of space, so as to provide sudden enlightened feelings for people.

The Space with Visual Extension Effect

延伸视觉的空间

针对人流量大的狭窄通道，可以设计有着直接视觉效果的大视角艺术图形。而针对空间大、人流速度慢的空间，比如地铁站的候车间，就可以在其细部添加艺术元素，让人仔细欣赏，消磨时间。除此以外，可利用人们视觉的特点，通过对空间界面如顶棚、地面、墙面的处理，玻璃、镜面等材料的应用，使空间产生开阔感和延伸感。例如在设计地下街时，巧妙地运用镜子的转折、反射效果，使空间印象充分扩张，会让人产生豁然开朗的感觉。

- **Architects:** 3XN A/S
- **Client:** Bygningsfonden Den Blå Planet
- **Location:** Kastrup, Copenhagen, Denmark
- **Gross Area:** App. 10,000m², where of app. 5,000m² exhibition. Outdoor area app. 2,000m² plus parking area for 200 vehicles, in total parking for 575 vehicles

- 设计公司：3XN A/S
- 客户：Bygningsfonden Den Blå Planet
- 地点：丹麦哥本哈根凯斯楚普
- 总面积：10 000m²，其中有 5 000 m² 的展览厅，室外面积 2 000 m²，加上能容纳 200 辆汽车的停车区域，总共能停车 575 辆。

"蓝色星球"水族馆

The Blue Planet

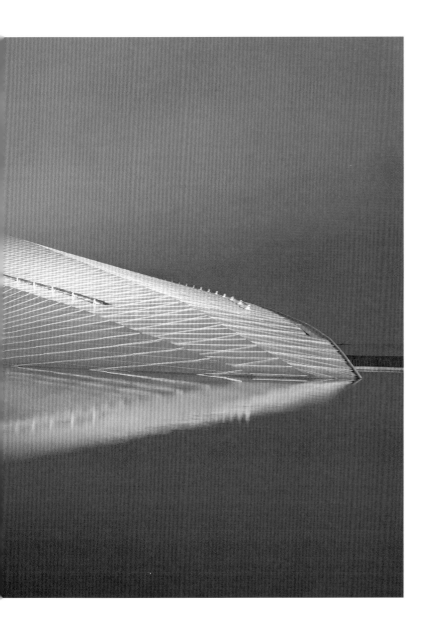

◯ Project Overview

Inspired by the shape of water in endless motion, Denmark's new National Aquarium, The Blue Planet is shaped as a great whirlpool. The Blue Planet is located on an elevated headland towards the sea, north of Kastrup Harbor. The building's distinctive shape is clearly visible for travelers arriving by plane to the nearby Copenhagen Airport.

◯ 项目概况

受到水无休止运动形式的启发,丹麦最新的自然水族馆蓝色星球被设计成一个巨大的旋涡状建筑。蓝色星球坐落在卡斯特鲁普港北部海岸上。游客乘飞机抵达哥本哈根机场附近时,一眼便可看见这个外观独特的建筑。

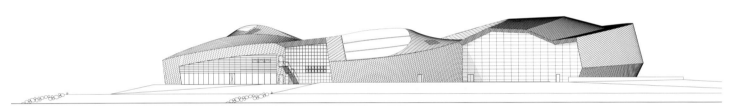

Northern Façade 北立面图

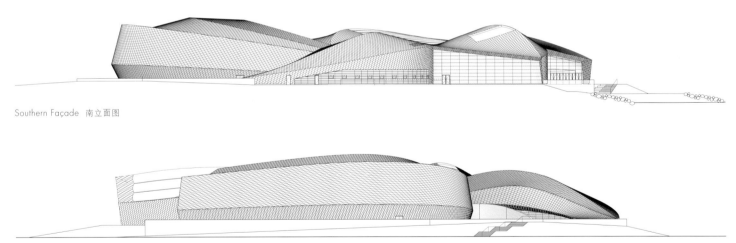

Southern Façade 南立面图

Western Façade 西立面图

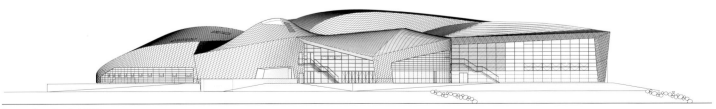

Eastern Façade 东立面图

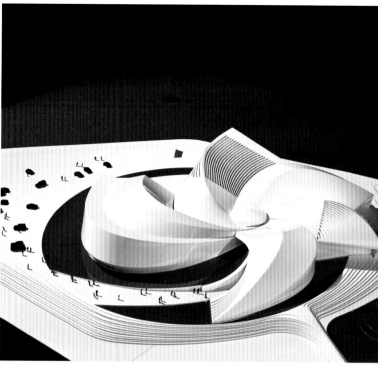
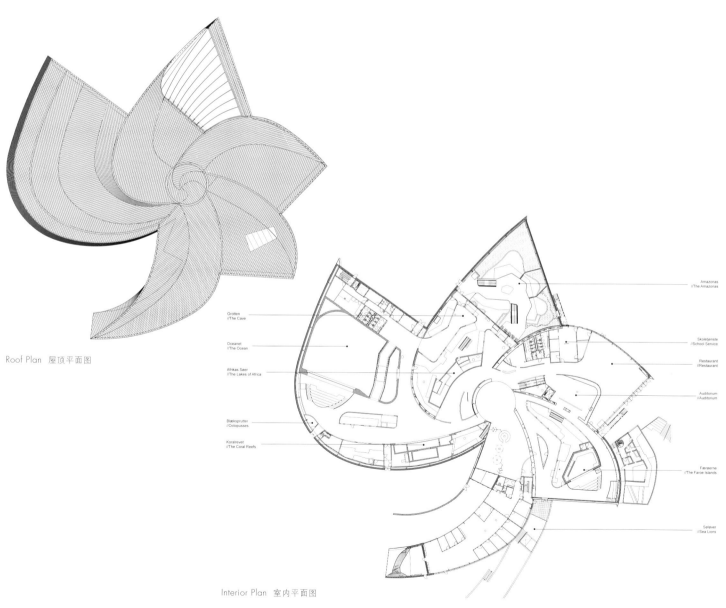

Roof Plan 屋顶平面图

Interior Plan 室内平面图

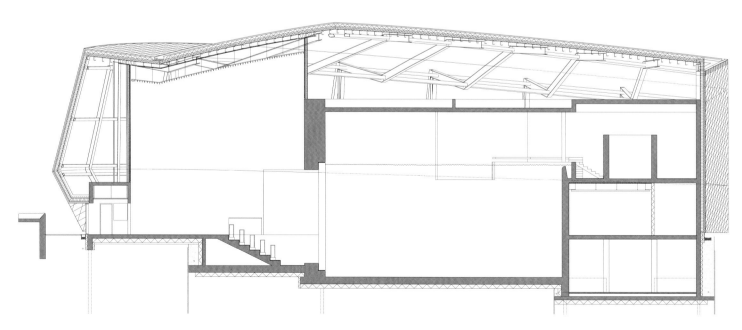

Section Detail 剖面细节图

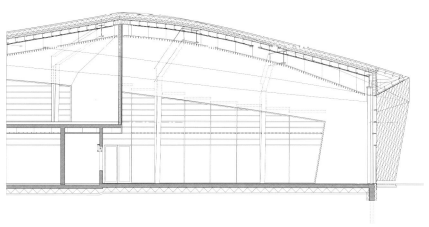

Section Detail 剖面细节图

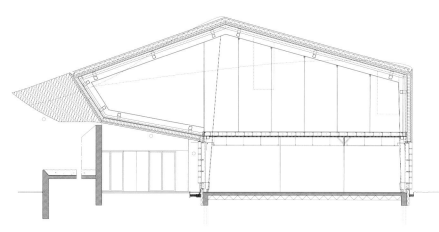

Section Detail 剖面细节图

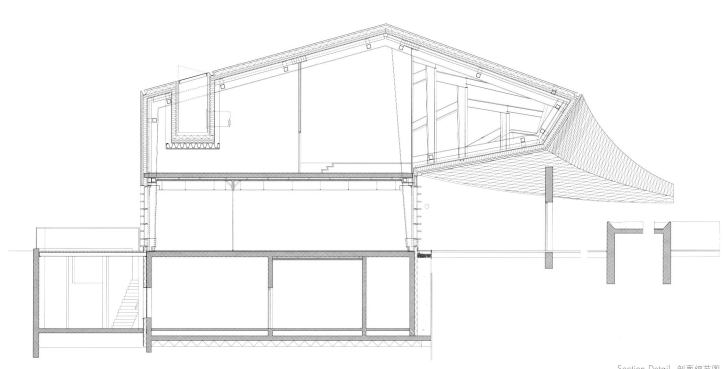

Section Detail 剖面细节图

307

○ **Design Highlights :The Whirlpool**

As an image, the whirlpool is at once both abstract and figurative. It stirs attention with its distinctive vortex blades, but at the same time, as a building, changes dramatically depending on viewing angle, distance and daylight conditions. From the air, almost entirely white, its contours are reminiscent of a starfish. From the front, the building's organic lines are evocative of silvery-grey waves or a vast sea creature, and on closer inspection, the façade patterning is reminiscent of fish scales. The facade is covered with more than 33,000 small diamond-shaped aluminum shingles, which adapts to the building's organic form.

○ **设计亮点：漩涡形的空间**

作为一个概念，旋涡是既抽象又具象的。它因为其独特的涡流形叶片而引起了人们的关注，同时，作为一栋建筑，其奇特的变化与观察的角度、距离和日光照射条件密切相关。从空中看，它几乎是白色的，它的轮廓看上去会让人想起海星。从前面看，建筑的有机线条就像海上唤起的银灰色波浪，又或者像一个巨大的海上生物。如果仔细去看的话，你会发现正面的图案会让人想起鱼鳞。建筑的正面被超过 33 000 小钻石形状的铝合金板覆盖，这些都适应了建筑的有机形式。

The whirlpool concept was chosen as ideal not only for its visual associations, but also because it resolved a practical challenge in the design brief: it ensures that one or more of the whirlpool arms, with relative ease and without disrupting the building's integrity nor the operation of the aquarium, can be extended with more than 30 % in order to create more exhibition space.

旋涡被选为理想的建筑模型，不仅在于它能引起人们形象化的联想，也在于它在设计概要中解决了一个具有挑战性的实际问题：它确保了一只或多只旋涡体的分肢能得到相应的舒展，并且不影响建筑的完整性以及水族馆的运作，可以增加超过30%的面积以便创造更多的展览空间。

Visitors reach the entrance by following the first and longest of the whirlpool's arms, already starting in the landscape. With a smooth transition the landscape surpasses for the building, while the outdoor ponds mark the unique experience that awaits the aquarium visitors as they enter: the whirlpool has pulled them into another world-a world beneath the surface of the sea.

游客沿着漩涡的第一个并且是最长的分肢到达入口处，在绿化带中开始了旅程。通过从绿化到建筑这一段平稳的过渡，户外的池塘标记了游客在等待进入蓝色星球时的独一无二的体验：漩涡将游客带到了另一个世界——一个水下世界。

A circular foyer is the center of motion around the aquarium, and it is here visitors choose which river, lake or ocean to explore. By enabling multiple routes the risk of queues in front of individual aquariums is reduced. The interiors range from grand to intimate settings, allowing the architecture and the exhibits to jointly convey an array of diverse environments and moods. The curved ceilings of the aquarium are reminiscent of the baleens of a large whale.

圆形大厅是水族馆的运动中心，也就是在这里，游客可以选择探秘河流、湖泊、大海。通过使用多个路径，在水族馆前排队的个人风险降低。在水族馆内部，无论是宏大的设计，还是与场馆紧密相关的设置，都使得建筑和参与展览的动物共同传达了环境的舒适度，以及动物的心情指数。弯曲的天花板看起来就像巨大的鲸的须。

The exhibition is a total concept offering all visitors a sensuous and captivating experience of life in and under the water. A mixture of light, sound, advanced AV-technology, projections, film, interactivity, graphics, illustrations and signs aimed at all age levels ensures that every visitor, regardless of background or interests, has the best experience possible. As the only aquarium in Denmark, The Blue Planet focuses on all aquatic life – from cold and warm waters, fresh and salt. In total, The Blue Planet contains app. 7 million liters of water and 53 aquariums and displays.

水族馆的展览为游客们提供了感性的、迷人的水下世界场景，提高了他们对水下世界的认识。光线、声音、投射、电影、现场互动、图片、说明和针对各个年龄层的标志巧妙地结合在一起，使背景不同和兴趣不同的游客，都能得到最好的体验。作为丹麦唯一的水族馆，蓝色星球水族馆非常关注所有水生动物的生活状况——无论是水的冷暖，水的新鲜度，还是盐的浓度，都经过严格的把关、控制。总的来说，蓝色星球水族馆共有 7 000 000 升水，53 个水族馆和展区。

The restaurant's decor is based on the colors and expressions that characterize Nordic nature. The restaurant faces south-east, and thus offers a panoramic view of the sea. The outdoors facilities include a terrace with seating, a pond with carps and a tank with sea lions. The sea lions can also be looked at from the inside of the aquarium.

水族馆的餐厅根据北欧自然风情和色调进行了装饰。餐厅面向东南，游客可从餐厅看到海的全景。户外的设施包括一个带有座位的阳台，一个养着鲤鱼的池塘和一个有海狮的水池。游客可以从水族馆里面看到海狮的身影。

◯ **Design Details**

The building extends beyond the original coastline, placing special requirements on the facility's structures In a terrain with tendency to subsidence. The structure is founded on piles and all of the sewage structures are suspended in the concrete structure. The building's architectural façade design forms the basis for the design of the steel structures. The load-bearing system consists of 54 unique steel frames, which via their radial positioning and geometry forms the base of the curved façade. A service line was built 1.7 km out into the Øresund to obtain suitable water for the aquariums. Moreover, the cooling system for aquariums and climate system for public areas also use seawater.

◯ 设计细节

建筑的延伸超越了原来的海岸线，对位于下沉式地形的设施结构有着特殊的要求。建筑是建立在悬浮在混凝土结构的桩和下水道之上的。建筑的立面设计构成了钢结构设计的基础。承重体系包含 54 个独特的钢结构，这些钢结构通过其径向定位和几何形状搭建起弧形外墙的基础。施工团队在深入厄勒海峡 1.7 km 的地方建了一条服务线路，专门为水族馆获取合适的水源。此外，水族馆的冷却系统和公共领域的气候调节系统也是使用海水的。

- **Architects:** ON-A
- **Client:** TMB (TRANSPORTS METROPOLITANS BARCELONA)
- **Photographer:** Lluis Ros
- **Location:** Barcelona, Spain
- **Floor Area:** 3,500m²

- 设计公司：ON-A
- 客户：TMB (TRANSPORTS METROPOLITANS BARCELONA)
- 摄影师：Lluis Ros
- 地点：西班牙巴塞罗那
- 占地面积：3 500m²

Sant Andreu 地铁站改造

Sant Andreu Metro Station Reform

◯ **Project Overview**

Can a subway station change its look? Can the passengers change the image of the space? We want the habitual user to perceive the station distinctly each day so that they may not be indifferent to their waiting time.

◯ 项目概况

一个地铁站可以改变它的外观吗?乘客们能改变对地铁空间的印象吗?我们就是要这些有着定向思维的乘客将注意力转移到一个独特的地铁空间,让他们不再觉得候车就是在无聊地打发时间。

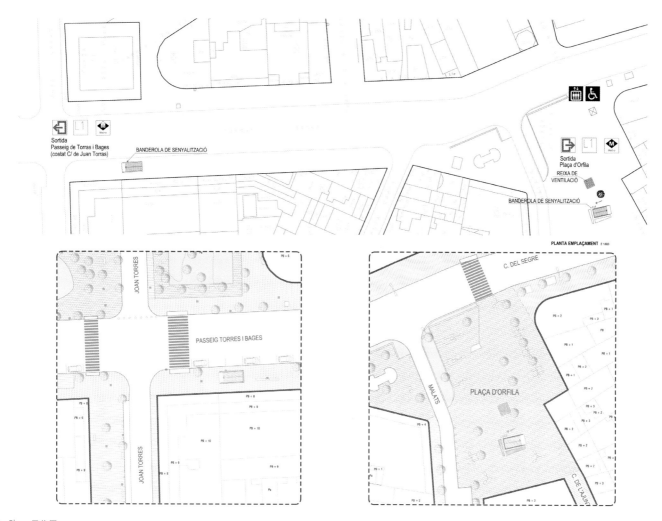

Site Plan 区位图

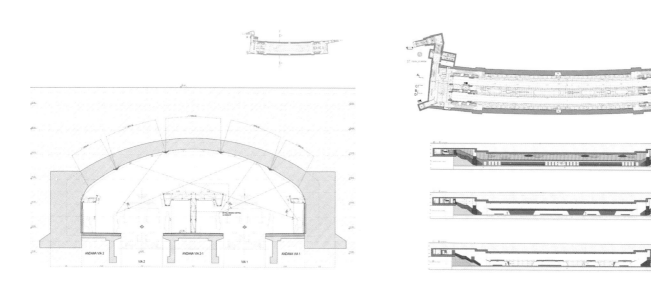

Cross Section 横向剖面图

Longitudinal Section 纵向剖面图

○ Design Highlights: Flexible and Alive Platform

The main intention of the refurbishment of the Sant Andreu's station was to renew the cladding, signage and facilities in the most serious and effective way possible to emphasize and highlight the intervention of the dome. Illuminating this great space is a complex system of lighting and projectors, which are located in the structure of the central platform, allowing for the creation of different scenes that make the area flexible and alive.

○ 设计亮点：灵动而鲜活的站台空间

Sant Andreu 地铁站的翻新计划是以严谨、有效的方式对室外装饰、标志牌和设施进行翻新，以强调和突出圆顶图案对地铁站的介入和影响。这个庞大空间的照明是通过一个复杂的采光和投影系统来实现的。这些系统位于站台空间的建筑当中，投射出不同的映像，令地铁空间变得灵动而鲜活起来，这让地铁站看起来显得更有创造性。

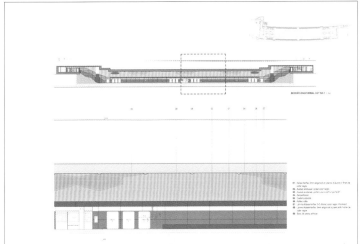

Longitudinal Section 纵向剖面图

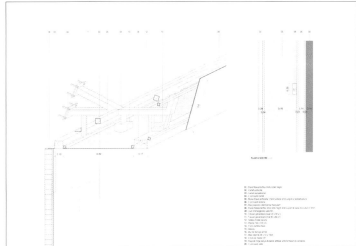

Axonometry 轴视图

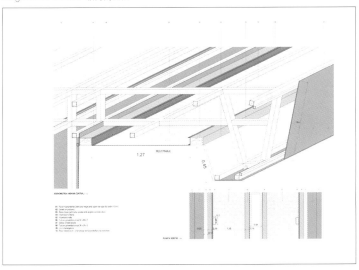

Axonometry 轴视图

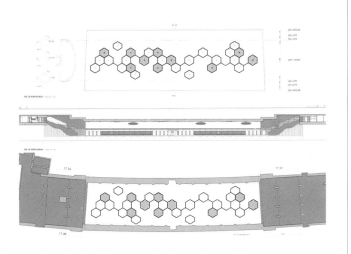

Axonometry 轴视图

317

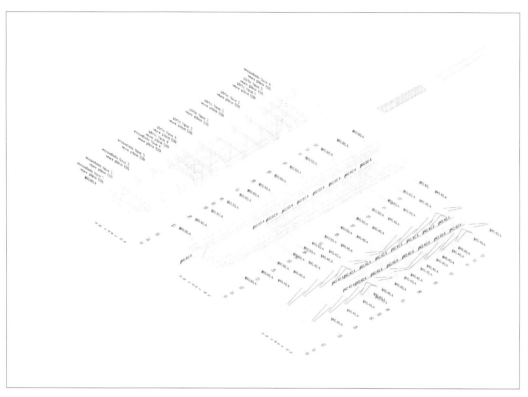
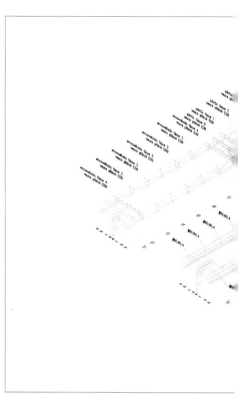

Axonometry 轴视图

Axonometry 轴视图

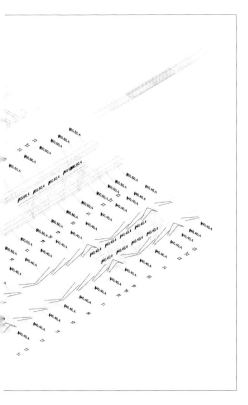

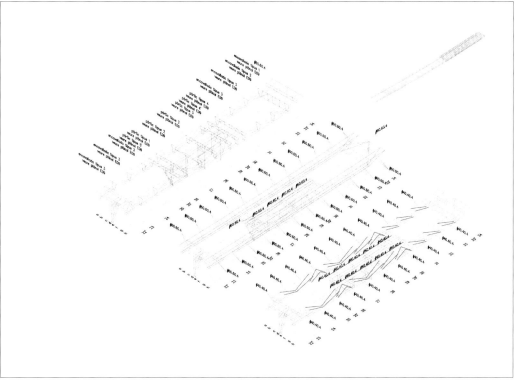

Axonometry 轴视图

- Architects: ON-A
- Client: TMB (TRANSPORTS METROPOLITANS BARCELONA)
- Photographer: Lluis Ros
- Location: Barcelona, Spain
- Floor Area: 1,500m²

- 设计公司：ON-A
- 客户：TMB (TRANSPORTS METROPOLITANS BARCELONA)
- 摄影师：Lluis Ros
- 地点：西班牙巴塞罗那
- 占地面积：1 500m²

Drassanes 地铁站改造

Drassanes Metro Station Reform

○ **Project Overview**

Drassanes Metro Station is located in one of the most popular city in Europe-Barcelona. Being similar to most of the metro station in Spain, it is built in the narrow and enclosed underground space. After reforming by the ON-A Architects, the metro station becomes fresher and brighter. And it provides a completely new sense to the passengers.

○ **项目概况**

Drassanes 地铁站位于欧洲最受欢迎的城市之———巴塞罗那。同西班牙大多数地铁站一样，它修建在狭窄的、封闭的地下。ON-A 建筑事务所对其进行了改建，使地铁站变得更加清新、明亮，让乘客有耳目一新的感觉。

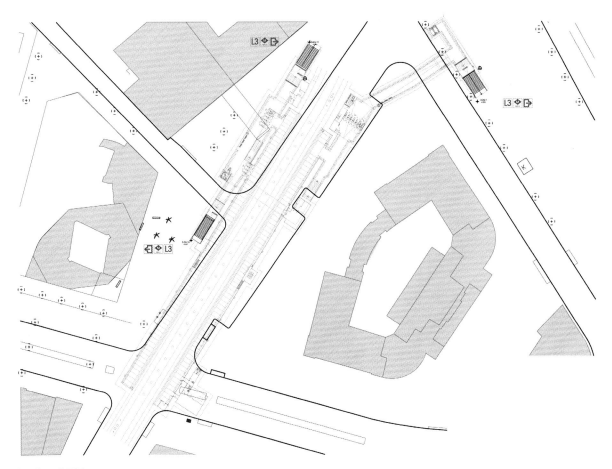

Site Plan 总平面图

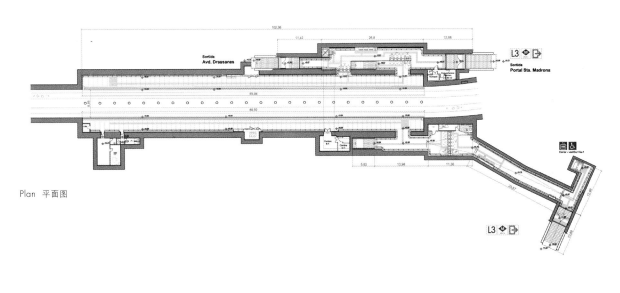

Plan 平面图

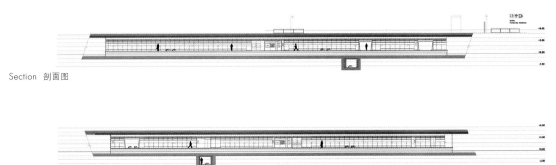

Section 剖面图

Section 剖面图

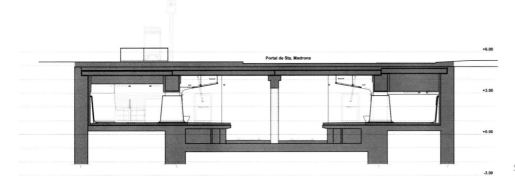

Section 剖面图

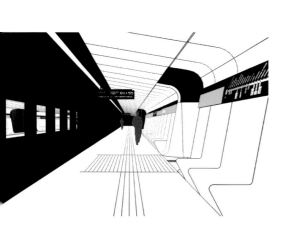

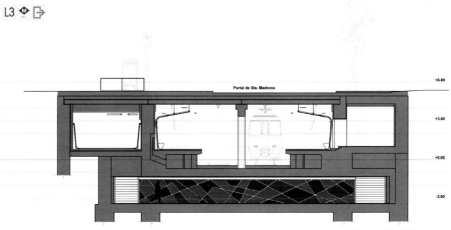

Section 剖面图

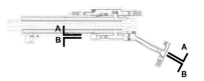

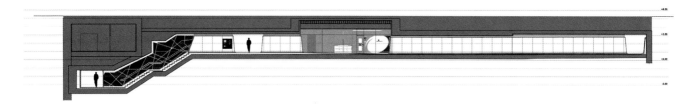

Section 剖面图

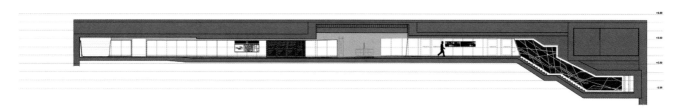

Section 剖面图

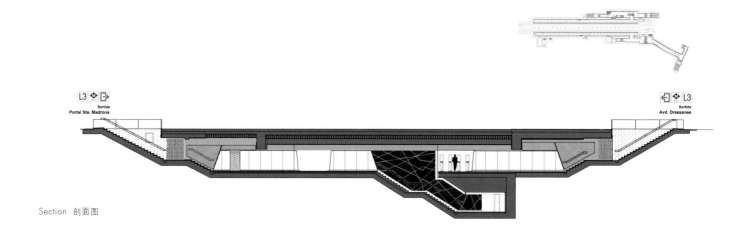

Section 剖面图

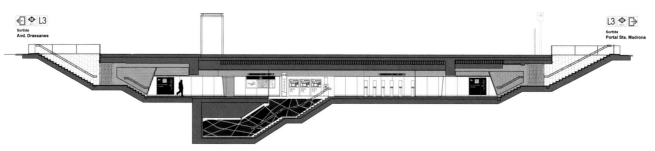

Section 剖面图

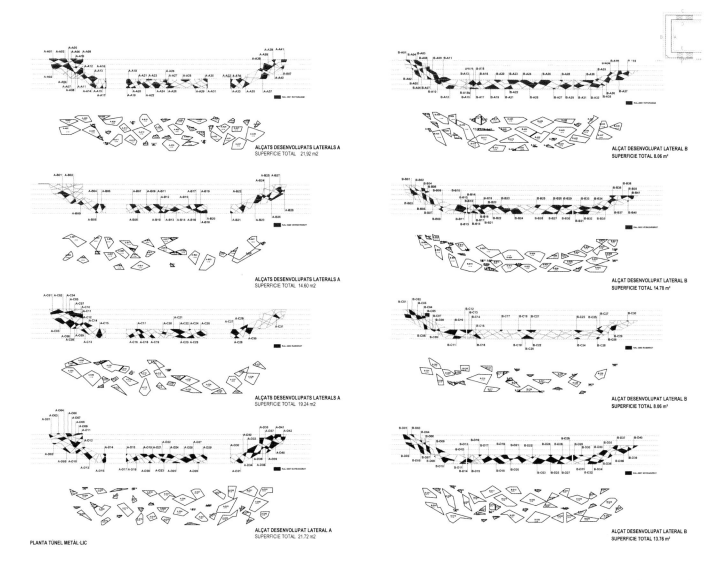

Design Concept 设计概念图

◯ Design Highlights : Unimpeded and Indusive Environment

There are two problems in the metro station reforming. Firstly, the space of the metro station is limited. The design concept must base on the existing system such as the ventilation system, the circuit system, water-heating system in order that the huge public infrastructure can still work in the reforming period as well as the existing structure can be preserved and the waste caused by city construction can be controlled. Secondly, a clear and inclusive environment is needed for the new metro station. They must solve the problems above in order that the design concept contains all the essential function elements of the metro station, and they also want to equip passengers with everything they need. But the structure and the size of the metro station can't be changed. So they can only change the interior decoration to reflect the innovative value of the reforming. They decided to reform the materials and the interior decoration of the metro station by following the interior decoration of the metro train so that there is a natural transition between the metro station and the metro train. Meanwhile, the designer use the glass reinforced concrete panels for construction because they want to create a bright and clear interior environment. For the tunnel part, the designers design an irregular red vitreous wall. The black floor, black ceiling and the red walls with extremely strong visual impact, feature the Spanish Characteristics.

◯ 设计亮点：通畅和包容的内部环境

在进行地铁站的改建时，设计师碰到两个问题：第一，空间非常有限，必须结合现有结构（比如通风、电路、水暖系统）进行设计。只有这样，才能保证这个庞大的公共基础设施在改建期间仍然能够正常运作，才能维持旧有的框架结构，减少城市建设中造成的浪费。第二，新地铁站需要一个通畅和具有包容度的内部环境。他们必须解决这两个问题，使整个设计具备地铁站应有的功能元素，并能够满足乘客所有的需要。由于该地铁站的空间结构和体量大小不能改变，改造的创新价值就只能从内部的装饰来实现。因此他们决定效仿地铁车厢的内部装饰，更新地铁站内部的材料和装饰，使车站和列车之间有一个自然的过渡。同时为了构造一个明亮、干净的室内环境，设计师选用了玻璃纤维增强混凝土面板。在地铁站通道部分，设计师设计了不规则的红色玻璃体墙面，与黑色的地面和天花与产生红色的墙体视觉冲击力，并富有西班牙特色。

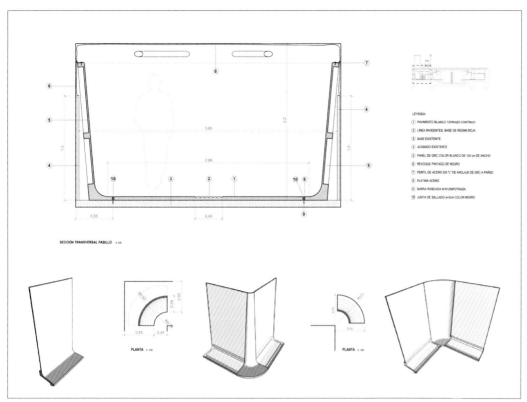

Detail 细节图

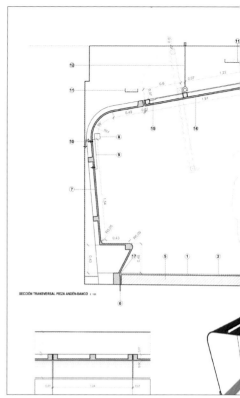

Detail 细节图

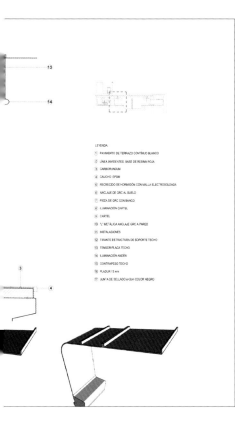
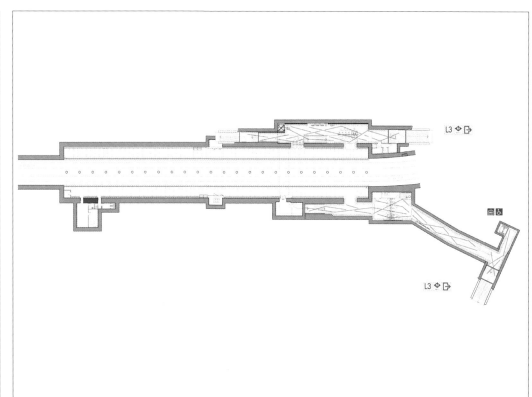

Plan 平面图

◯ Design Details

Finishing details that are similar to the interior of a train's freight car equip passengers with everything they need and provide a clean and bright appearance. When the passengers come to the station, their trails would weave together, paths crossing at the station. The lighting and the panels of the connecting tunnel reflect this idea of lives that intersect daily.

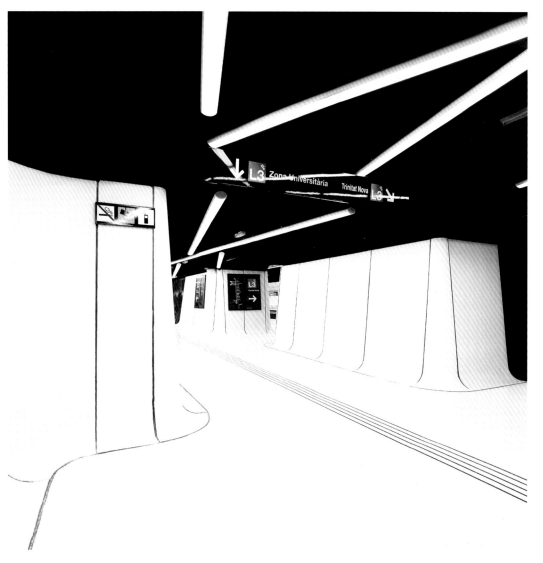

设计细节

从已完成的细节来看,这个地铁站很像一辆能满足乘客的各种需求的、有着整洁明亮外观的列车。当乘客来到地铁站,他们很快便会交汇在一起,尽管他们进入地铁站的路线并不一致。地铁站的采光系统和连接隧道的面板体现出这种日常人流交汇的设计思路。

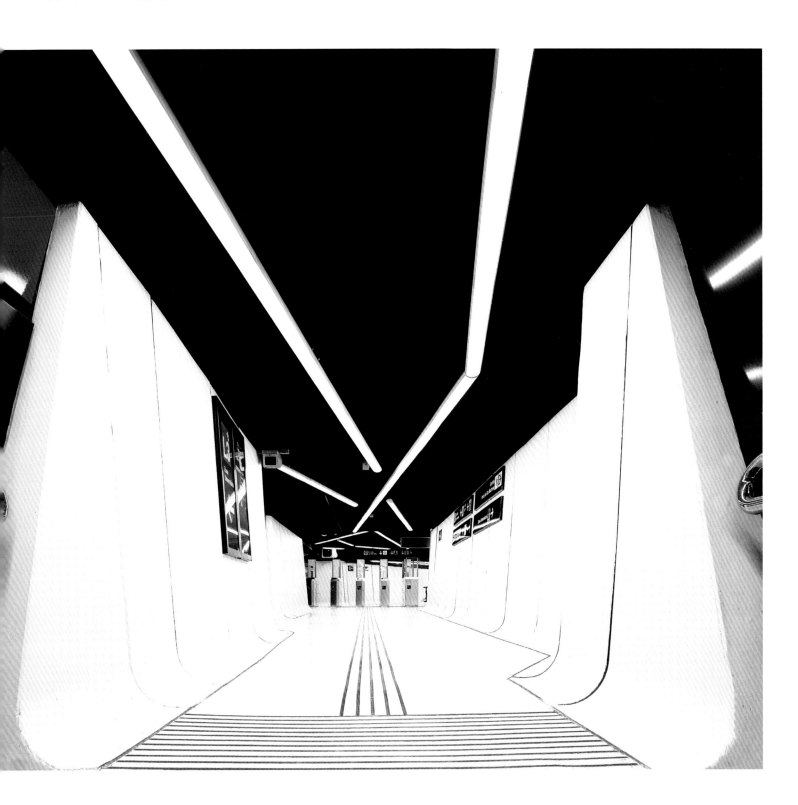

- **Architects:** Oscar Tusquets Blanca
- **Project Director:** Giovanni Fassanaro
- **Lighting:** AIA(project: Albert Salazar, collaborator :Pablo Martínez)
- **Artists:** William Kentridge, Bob Wilson, Achille Cevoli, Francesco Clemente, Ilya y Emilia Kabakov, Shirin Neshat y Oliviero Toscani,
- **Location:** Naples, Italy
- 设计公司：Oscar Tusquets Blanca
- 项目负责人：Giovanni Fassanaro
- 灯光设计：AIA 公司（项目负责：Albert Salazar，合作者：Pablo Martínez）
- 艺术设计：William Kentridge、Bob Wilson、Achille Cevoli、Francesco Clemente、Ilya y Emilia Kabakov、Shirin Neshat y Oliviero Toscani
- 地点：意大利那不勒斯

Toledo 地铁站

Toledo Metro Station

◯ **Project Overview**

Toledo Metro Station is located in the Naples, Italy. It is design by the artist Oscar Tusquets Blanca. Unlike traditional metro space, the overall design of Toledo Metro Station is not inflexible. The main color of the station is blue, which makes the metro space look like a realistic underwater world.

◯ 项目概况

Toledo 地铁站位于意大利的那不勒斯,由艺术家 Oscar Tusquets Blanca 倾心打造,整体设计打破了传统地铁空间的刻板,用扑面而来的蓝色将其缔造成一个不折不扣的海底世界。

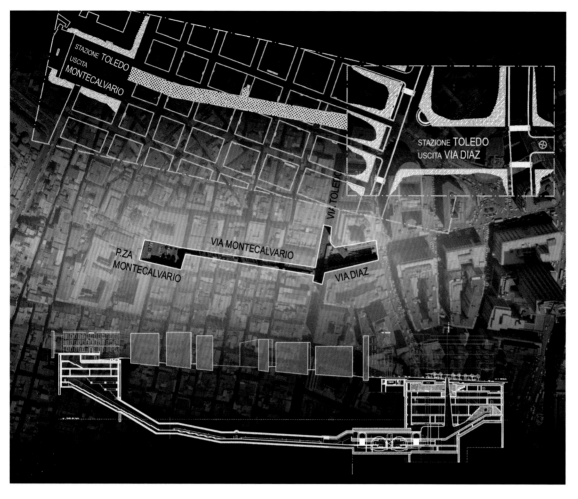

There is another exit in the Toledo Metro Station-the Montecalvario Exit. The designers fully consider the local environmental situation before building this exit. Montecalvario Square has turned into an outdoor living room. It is suitable for pedestrians and also to accommodate children from the facing school. But it is impossible to hollow the centre of the only little neighbourhood square with access stairs to the subway. So the designers decided to use the ground slope to make a completely horizontal space that would allow access almost at street level from the street below. The connection between the uneven streets and the horizontal square is

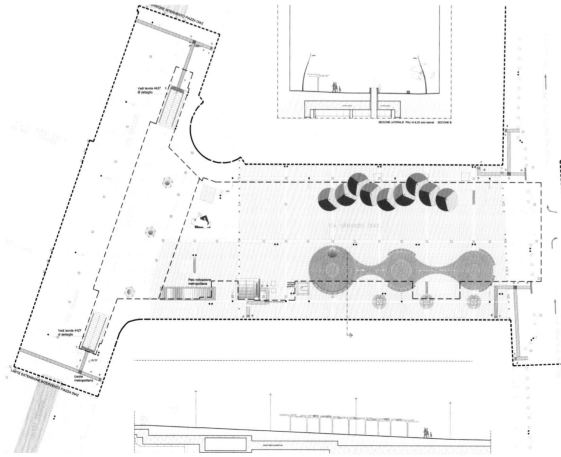

Plan 平面图

solved with very wide and comfortable stairs. We wanted to incorporate plants and trees, not only to generate shadow spaces but also to create a little "green space" in the Quartieri Spagnoli, where there is not vegetation at all. We can learn the unique anthropological context of the city of Napoli from the exit.

这个地铁站有另一个出口——Montecalvario广场出口。在建造这个出口前，设计师充分地考虑了当地的环境因素。目前，Montecalvario广场已经变成一个户外的客厅，它能容纳众多的行人以及来自对面学校的孩子。但是把这个邻近的小广场的中心挖空，建造连接地铁的出入口和楼梯，这是不可能的。所以设计师决定利用地面的斜坡，创造了一个完全平整的、能够很好连接地上、地下的空间。

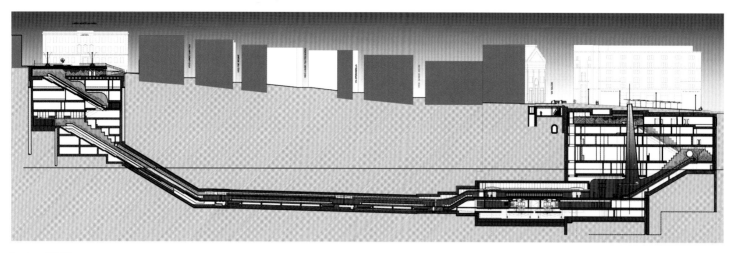

Section 剖面图

为了解决高低不平的街道与这个完全平整的广场的连接问题，设计师建造了宽阔、舒适的楼梯。通过这个出口，我们可以领略到那不勒斯市独特的文化背景。

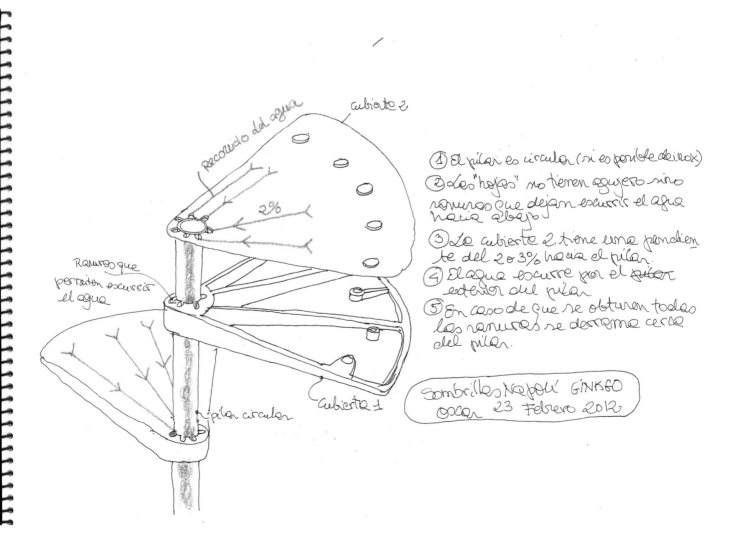

○ **Design Highlights : The Wall with Two Different Styles**

The other driving idea arose because much of the station is below sea level. The parts above sea level seem to be excavated in the rock whilst the lower levels seem sunk in the sea. The floor and walls of the upper part are excavated+ in natural stone, and the appearance of the aragonese wall in the ticket hall heightened the image of excavation. The "underwater" areas are entirely covered with blue vitreous mosaic. Above, everything is earthy and matte; below, blue, shining and vibrant.

○ 设计亮点：一分为二的墙壁

地铁站的大部分地方都位于海平面的下面，这使得设计师产生其他富有建设性的想法。地铁站在海平面之上的部分看起来是从岩石中挖掘的，而其海拔较低的部分看起来像是沉在海底的。上部的地板和墙壁都是用天然的石头砌成的。售票厅中有一堵富有阿拉贡地区特色的墙，这堵墙强化了挖掘工程的图像效果。"水下"的区域，全部铺上蓝色玻璃马赛克。地上的部分，都是用土质材料建造的，是不光滑的；地下的部分，是蓝色的、闪亮的，充满了生气。

This dichotomy is heightened by the works of two artists who is cooperated with Oscar. In the "excavated" areas William Kentrid.ge made two murals in stony mosaic, with a rough, ancient character. In the "aquatic" areas Bob Wilson installed, facing one another, two enormously long views of the seashore whose waves move subtly as one walks along the gallery. Wilson also played a fundamental role in lighting the crater, creating a programme to coordinate the colour and intensity of each LED over time.

这种一分为二的设计方法，在两位和 Oscar 合作的艺术大师那里得到强化。在地上的部分，William Kentrid.ge 在石制的马赛克上创作出两幅表面粗糙的、有着古典特色的壁画。在地下的部分，Bob Wilson 沿着过道，制作了两个很长的、面对面的动态图像。当游客走过这里的时候，他们可以看到翻腾的海浪拍打着岸边。Wilson 也做了不少举足轻重的工作，无论是地铁的深坑的照明，还是 LED 图像的颜色、光线强度的调和，都做得非常到位。

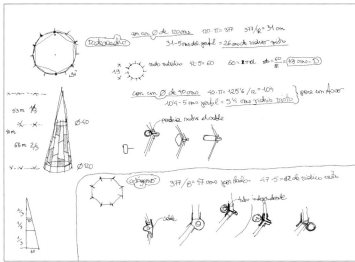

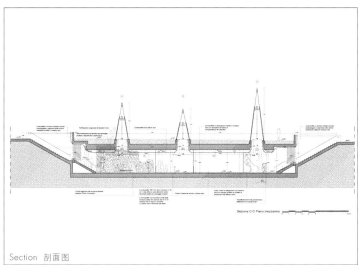

Section 剖面图

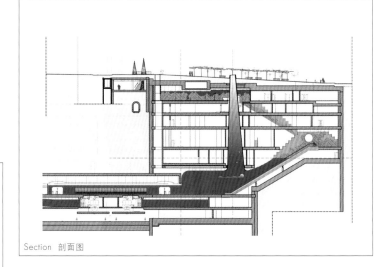

Section 剖面图

◯ Design Details

The leitmotiv of the project emerged when the works had barely started, making substantial structural modifications possible. The pit, more than 40 m deep, beneath the crane that was extracting tonnes of volcanic sediments, was Piranesian in its grandeur. It was intended to cover this cavern when the crane had finished, but finally the cavern is preserved. Because it can make travelers sense how deep they were and travelers can glimpse the sunlight above, while in the piazza strollers could lean over the parapet and see passengers moving around 37 m below, something surprising and dizzying.

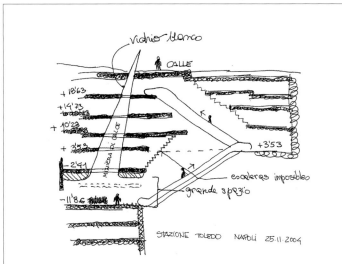

○ 设计细节

工作才刚刚开始，项目的主题就呈现出来了，这使得结构上的实质性变化得以实现。在施工过程中，在提取数以吨计的火山沉积物的起重机的下面，有一个深度超过 40 m 的坑。这个坑带有皮拉内西式的宏伟风格。本来打算在起重机完成挖掘后就把这个坑填上，但最后还是把这个坑保留下来了。因为这可以让旅客感受到他们处于多么深的位置上，让他们在无意的一瞥中，看到上面的阳光。而在上面广场散步的行人可以依着栏杆，看到下面 37 m 的乘客在移动，也能看到一些令人惊喜的、令人眼花缭乱的东西。

Apart from through the great crater, natural light penetrates the mezzanine through three skylights that illuminate the interior while also enabling passers-by to glimpse the aragonese wall and Kentridge's murals.

阳光除了可以照进巨大的深坑，还能够透过夹层，通过三个天窗照进室内，这样乘客也能看见里面的有着阿拉贡地区风格的墙和 Kentrid 创作的壁画。

Index 索引

3XN A/S

The aim of the companyis to create architecture for people through a complex approach tailored to users and clients. The designers build on the Scandinavian tradition of clear functionality and combine a broad knowledge of materials, technology and user needs with organizational strengths in financial management, project management and execution.

公司的目标是根据用户或客户的要求，为他们创建合适的建筑。设计师结合斯堪的纳维亚的传统，建造具有清晰的功能和简约之美的建筑。他们能够将对材料、技术和用户需求的广泛认知与财务管理、项目管理和执行方面的优势很好地结合在一起。

Anttinen Oiva Architects

Based in Helskini, Anttinen Oiva Architects was founded in 2006 by Selina Anttinen and Vesa Oiva. The Company's aim is to create ambitious, memorable and versatile contemporary architecture, which takes its inspiration from the demanding context.

Anttinen Oiva 建筑事务所位于赫尔辛基，在 2006 年由 Selina Anttinen 和 Vesa Oiva 创立。公司的目标是从客户需求出发，激发灵感，创造规模宏大的、令人难忘的和灵活的现代建筑。

BNKR Arquitectura

Bunker Arquitectura is a Mexico City-based architecture. In their short career they have been able to experience and experiment architecture in the broadest scale possible: from small iconic chapels for private clients to a master plan for an entire city.

BNKR 建筑事务所是一家位于墨西哥城，并根据墨西哥城的特点进行建筑设计、城市规划和研究的公司。在短短几年间，他们就已经具备丰富的经验和把建筑应用到最广阔的领域的能力：他们的工作范畴也从建造小型的标志性的教堂，扩展到为整个城市规划方案。

BIG

BIG is a Copenhagen and New York based group of architects, designers, builders and thinkers operating within the fields of architecture, urbanism, research and development.

BIG 建筑事务所在哥本哈根和纽约都有自己的建筑、设计、建造和规划团队，所从事的工作涉及建筑、城市化、研究和开发领域。

C.F. Møller

C.F. Møller is one of Scandinavia's oldest and largest architectural practices. Their award-winning work involves a wide range of expertise that covers all architectural services, landscape architecture, product design, healthcare planning and management advice on user consultation, change management, space planning, logistics, client consultancy and organisational development.

C.F. Møller 是斯堪的纳维亚历史最悠久的、规模最大的建筑公司之一。他们的工作能体现出很强的专业性，曾多次获奖。他们可以在多个领域为客户提供服务，包括建筑、景观、产品设计、环保规划与客户管理咨询、变革管理、空间规划、物流、客户咨询与建筑开发。

CL&AA

Claudio Lucchin & Associates practice was founded in 2004 when Lucchin's closest collaborators Angelo Rinaldo and Daniela Varnier joined the studio. They're involved with architectural and urban design. In 2013 they were selected for the Alto Adige Architecture Prize.

Claudio Lucchin & Associates practice 工作室成立于2004年，Lucchin 最亲密的合作者 Angelo Rinaldo 和 Daniela Varnier 在这一年加入了工作室。他们的业务范围包括建筑和城市设计。在2013年，他们获得了 Alto Adige 建筑奖。

Collingridge & Smith Architects

Collingridge And Smith Architects (UK) Ltd. is an award winning international architectural practice, founded by Graham Collingridge and Phil Smith in 2012, delivering unique, innovative and sustainable architecture.

Collingridge And Smith 建筑有限公司（英国）是一家屡获殊荣的国际建筑公司，它是由 Graham Collingridge 和 Phil Smith 在 2012 年建立的。他们提倡打造独特的、有创新性的、适合可持续发展的建筑。

Söhne & Partner Architekten

The architecture of the company is always tailored to the needs of the people who live, work or spend their leisure time in their buildings. They take a structured approach to problems, and solve them creatively.

这家建筑公司会根据人们的居住、工作或进行休闲娱乐的需要进行建筑设计。设计师根据问题提出结构方案，往往发挥他们的潜力想出最佳方案，以解决问题、满足客户需求。

Gonzalo Vaíllo Martínez

Gonzalo is an excellent architecture student of University of Alcala ETSAG, which is in Spain. He got A for The Cave as the Degree Project. The project of this public building began during the time they have an exchange program with the Bartlett School of Architecture in London.

Gonzalo 是西班牙的高校——阿尔卡拉大学的建筑学学生。他在毕业考试中，凭着项目 The cave 取得了 A 级的好成绩。这个项目是在他们与伦敦的 Bartlett 建筑学校交流的期间开始进行的。

John McAslan + Partners

John McAslan + Partners is a leading architectural practice based in London, with offices in Manchester, Edinburgh and Doha. An extensive portfolio of international award-winning projects includes infrastructure, hospitality, commercial, residential, education, cultural, heritage, urban design, and landscape sectors.

John McAslan + Partners 是一家位于伦敦的领先的建筑公司，在曼切斯特、爱丁堡和多哈都设有办事处。公司设计了不少得奖项目，项目的范围包括基础设施、酒店、商业、住宅、教育、文化、遗产、城市设计、景观。

Karres en Brands LR

Karres and Brands is an international design office for landscape architecture and urban planning. The office was founded in 1997 by Sylvia Karres and Bart Brands, and works on a very wide range of projects, studies and design competitions in the Netherlands and abroad.

Karres and Brands 是一家景观建筑设计和城市规划的国际公司。公司由 Sylvia Karres 和 Bart Brands 在 1977 年创建，在荷兰和海外业务面很广，包括项目设计、研究和设计竞赛。

Spektrum Arkitekter ApS

Spektrum Arkitekter is a young architectural practice that works with urban planning, building and landscape design and ties these disciplines inextricably together.

Spektrum 建筑事务所是一家有活力的公司，致力于城市规划、建筑与景观设计，并把这些工作有机地结合起来。

Kenji Mantani Studio

Kenji Mantani Studio providing the professional Japanese architecture design ideal to the client, is focused on the commercial real estate. The studio successfully transformed the commercial facilities by using the experience of transforming the creative industrial park in recent years.

万谷建筑设计公司提供专业的日本建筑设计理念，专注于商业地产设计。近年来，万谷活用改建设计创意园的丰富的经验来改建商业设施，取得极大的成果。

LAVA

LAVA explores frontiers that merge future technologies with the patterns of organisation found in nature and believes this will result in a smarter, friendlier, more socially and environmentally responsible future.

LAVA建筑事务所在合作者和组织者在自然界中共同发现的未来技术领域上进行探索。他们相信这样会造就一个更小的、更友善的、能平衡社会环境和自然环境的未来世界。

ON-A

ON-A, located in Barcelona since 2005, was created as an international architectural laboratory with great deal of professional experience. The main objective of the studio is to help give meaning to the discipline of architecture by creating interesting solutions, and developing these solutions with the utmost quality and respect to the design, technology and knowledge.

ON-A工作室位于巴塞罗那，从2005年到现在，已经打造出一个有着丰富经验的国际化建筑设计实验室。工作室的目的是根据每天的不同情况，提出富有创造性和趣味性的解决方案，以便在结合建筑原则的基础上进行设计，并在充分尊重特定环境下的设计、技术和知识背景的前提下，优化解决方案。

Zechner & Zechner ZT GmbH

Martin Zechner and Christoph Zechner run their main office with 30 architects in Vienna. Zechner & Zechner's engagement in urban design and large-scale buildings led to several commissions abroad.

Martin Zechner和Christoph Zechner在维也纳建立公司，他们的公司包含一个主要的办公室和30个建筑事务所。Zechner & Zechner公司很好地处理了城市规划和建造大型的建筑这两者的关系，为国外多个建筑委员会作了很好的引导。

Oscar Tusquets Blanca

Architect by profession, designer by adaptation, painter by vocation and writer out of the need to make friends, Oscar Tusquets Blanca is the prototype of the all-round artist that the specialisation of the modern world has brought to the verge of extinction.

Oscar Tusquets Blanca 有着专业的设计，包容的态度，有职业追求的画师，专注于工作的创作人员，所有艺术家的原型都可以在这里找到，这是一家有着当代世界少有的专业精神的设计公司。

Paolo Venturella Architect

Paolo Venturella Architecture, a recently open firm based in Rome, is involved in the research of different aspects of the city at different scale and works. The main analysis focus on contextualization, public space, sun exposure and sustainable renewable energies.

Paolo Venturella Architecture，是罗马一家新建筑公司，专注研究城市里各种规模和建筑的各方各面，重点研究建筑环境、公共空间、阳光照射和可持续再生能源。

Patkau Architects

Patkau Architects is an innovative architecture and design research studio based in Vancouver, Canada. In over 30 years of practice, Patkau Architects has been responsible for the design of a wide variety of building types.

Patkau 建筑事务所是一个富有创新性的建筑和设计研究的工作室，位于加拿大温哥华。经过 30 多年的实践，在加拿大和美国，Patkau 建筑事务所一直负责着为不同类型的客户设计多种多样的建筑。

Peter Ruge Architekten GmbH

Peter Ruge Architekten is an international architectural and urban design firm. For Peter Ruge Architekten a wide range of prospects has opened up concerning sustainable optimization of existing portfolios, residential developments and city planning.

Peter Ruge 建筑事务所是一个国际化的建筑和城市规划公司。Peter Ruge 公司在可持续发展优化、住宅开发和城市规划方面开辟了广阔的前景。

schneider+schumacher

The office schneider+schumacher Planungsgesellschaft mbH covers all stages of work, from initial design through to detailed design and overall planning. The emphasis is on detailed design, tendering, procurement, site supervision and property management, as well as quality control, and programming and cost control.

schneider + schumacher Planungsgesellschaft mbH 公司的业务范围涵盖设计工作的各个方面，从最初的设计，到细节方面的设计，再到整体的规划，都体现出设计团队的心血。公司工作的重点是细节设计、招标、采购、现场监督和物业管理，同时兼顾质量监控，编程和成本预算。

SOLDEVILA SOLDEVILA SOLDEVILA ARQUITECTES S.L.P.

The SOLDEVILA SOLDEVILA SOLDEVILA ARQUITECTES S.L.P. studio started in 2008 after the merger of 3 different studios. After the project for three L9-10 subway (Metro) stations, the group starts working in an integrated manner and in a single headquarters.

The SOLDEVILA SOLDEVILA SOLDEVILA ARQUITECTES S.L.P. 于2008年正式组建。它的前身分别为三家建筑工作室，在完成共同合作的三个L9-10地铁站项目后，这个团队开始组建在一起并成立一个总部。

SPORA ARCHITECTS

Spora architects is a Budapest based office - architects, designers, and thinkers operating within the fields of architecture, urbanism, research and development. The four partners, Tibor Dékány, Sándor Finta (till 2013), Ádám Hatvani and Orsolya Vadász opened practice in 2002. All of them graduated from the Budapest University of Technology and Economics. Among others, they currently work on one of Hungary's most significant projects, the construction of the underground line called Metro4.

Spoora 建筑事务所位于布达佩斯，拥有建筑师、设计师和规划人员团队，业务范围包括建筑设计，城市规划，研究和开发。有四个合作伙伴，分别是Tibor Dékány, Sándor Finta（一直到2013），dám Hatvani 和在2002年开公司的Orsolya Vadász。他们都毕业于布达佩斯技术与经济大学。他们目前的工作包括匈牙利的一个最著名的项目，地铁4号线的施工。

图书在版编目（CIP）数据

城市地下空间 / 凤凰空间·华南编辑部编. -- 南京：江苏凤凰科学技术出版社，2018.1
 ISBN 978-7-5537-8588-2

Ⅰ. ①城… Ⅱ. ①凤… Ⅲ. ①地下建筑物－建筑设计－作品集－世界 Ⅳ. ①TU92

中国版本图书馆CIP数据核字(2017)第248905号

城市地下空间

编　者	凤凰空间·华南编辑部
项目策划	韩　璇
责任编辑	刘屹立　赵　研
特约编辑	刘紫君
出版发行	江苏凤凰科学技术出版社
出版社地址	南京市湖南路1号A楼，邮编：210009
出版社网址	http：//www.pspress.cn
总经销	天津凤凰空间文化传媒有限公司
总经销网址	http：//www.ifengspace.cn
印　刷	北京科信印刷有限公司
开　本	965 mm×1 270 mm　1/16
印　张	22
字　数	176 000
版　次	2018年1月第1版
印　次	2018年1月第1次印刷
标准书号	ISBN 978-7-5537-8588-2
定　价	358.00元（精）

图书如有印装质量问题，可随时向销售部调换（电话：022-87893668）。